UBIQUITY

Ubiquity

THE SCIENCE OF HISTORY . . .

OR WHY THE WORLD IS

SIMPLER THAN

WE THINK

MARK BUCHANAN

CROWN PUBLISHERS
New York

Published by Crown Publishers, New York, New York.
Member of the Crown Publishing Group.

Random House, Inc. New York, Toronto, London, Sydney, Auckland
www.randomhouse.com

CROWN is a trademark and the Crown colophon is a registered trademark of Random House, Inc.

Originally published in different form, in Great Britain, by Weidenfeld & Nicolson in 2000.

Printed in the United States of America

Design by Karen Minster

Library of Congress Cataloging-in-Publication Data
Buchanan, Mark.
Ubiquity / Mark Buchanan.
p.   cm.
Includes bibliographical references and index.
1. Causality (Physics) 2. Pattern formation (Physical sciences) I. Title.
QC6.4.C3 B83 2001
530'.01—dc21   2001028365

ISBN 0-609-60810-X

10 9 8 7 6 5 4 3 2 1

First American Edition

To Kate, Nietzsche, and the Rabbit

Acknowledgments

✦

I am grateful to the many scientists who have taken the time to read and offer comments on portions of this manuscript, and to others who have generously supplied me with figures, data, or useful bits of information. My thanks go out especially to Robert Geller, Christopher Scholtz, Lena Oddershede, Tom Witten, Paul Meakin, Jim Meiss, Kim Christensen, Stefano Zapperi, Leo Kadanoff, Dietrich Stauffer, James Binney, Mark Newman, Bruce Malamud, Jeffrey Lockwood, Roy Anderson, Michael Benton, William Clemens, Peter Yodzis, Gene Stanley, Thomas Lux, Parameswaran Gopikrishnan, Dirk Helbing, Sidney Redner, Gottfried Mayer-Kress, and Lewis Miller. No one should suppose, of course, that any of these researchers agree with everything or even most of what I have written here. Any errors or misconceptions that may have slipped into the text are mine alone.

I would also like to thank Peter Tallack at Weidenfeld & Nicolson and Emily Loose at Random House for their faith in the value of this project, and, most important, my wife Kate for her unflagging support and encouragement during the many months of writing.

Mark Buchanan

Contents

✦

UBIQUITY

· 1 ·

Causa Prima

Politics is not the art of the possible.
It consists in choosing between
the disastrous and the unpalatable.

—JOHN KENNETH GALBRAITH[1]

———————

History is the science
of things which are never repeated.

—PAUL VALÉRY[2]

◆

IT WAS 11 A.M. ON A FINE SUMMER MORNING IN SARAJEVO, JUNE 28, 1914, when the driver of an automobile carrying two passengers made a wrong turn. The car was not supposed to leave the main street, and yet it did, pulling up into a narrow passageway with no escape. It was an unremarkable mistake, easy enough to make in the crowded, dusty streets. But this mistake, made on this day and by this driver, would disrupt hundreds of millions of lives, and alter the course of world history.

The automobile stopped directly in front of a nineteen-year-old Bosnian Serb student, Gavrilo Princip. A member of the Serbian terrorist organization Black Hand, Princip couldn't believe his luck. Striding forward, he reached the carriage. He drew a small pistol from his pocket. Pointed it. Pulled the trigger twice. Within thirty minutes, the Austro-Hungarian Archduke Franz Ferdinand and his wife Sophie, the carriage's passengers, were dead. Within hours, the political fabric of Europe had begun to unravel.

In the days that followed, Austria used the assassination as an excuse to begin planning an invasion of Serbia. Russia guaranteed protection to the Serbs, while Germany, in turn, offered to intercede on Austria's behalf should Russia become involved. Within just thirty days, this chain reaction of international threats and promises had mobilized vast armies and tied Austria, Russia, Germany, France, Britain, and Turkey into a deadly knot. When the First World War ended five years later, ten million lay dead. Europe fell into an uncomfortable quiet that lasted twenty years, and then the Second World War claimed another thirty million. In just three decades, the world had suffered two engulfing cataclysms. Why? Was it all due to a chauffeur's mistake?

On the matter of the causes and origins of the First World War, of course, almost nothing has been left unsaid. If Princip touched things off, to the British historian A.J.P. Taylor the war was really the consequence of railway timetables, which locked nations into a sequence of military preparations and war declarations from which there was no escape. The belligerent states, as he saw it, "were trapped by the ingenuity of their preparations."[3] Other historians point simply to German aggression and national desire for expansion, and suggest that the war was inevitable once Germany had become unified under Bismarck a half century earlier. The number of specific causes proposed is not much smaller than the number of historians who have considered the issue, and even today major new works on the topic appear frequently.[4] It is worth keeping in mind, of course, that all this historical "explanation" has arrived well *after* the fact.

In considering how well we understand the natural rhythms of human history, and in judging how able we are nowadays to perceive even the rough outlines of the future, it is also worth remembering that the century preceding 1914 had been like a long peaceful afternoon in European history, and that to historians of the time the wars seemed to erupt like terrifying and inexplicable storms in a cloudless sky. "All the spawn of hell," the American historian Clarence Alvord wrote after the First World War, "roamed at will over the world and made of it a shambles. . . . The pretty edifice of . . . history, which had been designed and built by my contemporaries, was rent asunder. . . . The meaning we historians had read into history was false, cruelly false."[5] Alvord and other historians thought they had discerned legitimate patterns in the past, and had convinced themselves that modern human history would unfold gradually along more or less rational lines. Instead, the future seemed to lie in the hands of bewildering, even malicious forces, preparing unimaginable catastrophes in the dark.

The First World War, the war sparked by "the most famous wrong turning in history,"[6] is the archetypal example of an unanticipated upheaval in world history, and one might optimistically suppose that such an exceptional case is never likely to be repeated. With the aid of hindsight, many historians now believe they understand the larger forces that caused the world wars of the twentieth century, and that we can once again see ahead with clear vision. But Alvord and his colleagues had similar confidence a century ago. What's more, few of us—professional historians included—seem any wiser when it comes to the present.

In the mid-1980s, the Union of Soviet Socialist Republics had existed for nearly three-quarters of a century, and it stood as a seemingly permanent fixture on the world stage. At that time, there were palpable fears in the United States that the U.S.S.R. was way ahead militarily, and that only with a concerted effort could the United States even stay competitive. In 1987, one would have had to scour the journals of history and political science to find even a tentative suggestion that the U.S.S.R. might collapse within half a century, let alone in the coming decade. Then, to everyone's amazement, the unthinkable became a reality—in just a few years.

In the wake of the U.S.S.R.'s unraveling, some historians leaped to another conclusion. Democracy seemed to be spreading over the globe, binding it up into one peaceful and lasting New World Order—the phrase favored, at least, by politicians in the West, who happily proclaimed the final victory of democracy (and capitalism) over communism. Some writers even speculated that we might be approaching "the end of history,"[7] as the world seemed to be settling into some ultimate equilibrium of global democracy, the end result of a centuries-long struggle for the realization of a deep human longing for individual dignity. Just a few years later, in what was then Yugoslavia, war and terrible inhumanity once again visited Europe. A momentary setback? Or the first ominous sign of things to come?

No doubt historians can also explain quite convincingly—though in retrospect, of course—why these events unfolded as they did. And there is nothing wrong with this kind of explanation; it is in the very nature of history that thinking and explanation must always proceed backwards. "Life is understood backwards," as Søren Kierkegaard once expressed the dilemma, "but must be lived forwards." And yet this need to resort always to explanations *after the fact* also underlines the seeming lack of any simple and understandable patterns in human affairs. In human history, the next dramatic episode, the next great upheaval, seems always to be lurking just around the corner. So despite their aim to find at least some meaningful patterns in history, it is probably true that many historians sympathize with the historian H. A. L. Fisher, who in 1935 concluded:

> Men wiser and more learned than I have discerned in history a plot, a rhythm, a predetermined pattern. These harmonies are concealed from me. I can see only one emergency following upon another . . . and only one safe rule for the historian: that he should recognize in the development of human destinies the play of the contingent and the unforeseen. . . . The ground gained by one generation may be lost by the next.[8]

Having read this far, you may be surprised to learn that this book is about ideas that find their origin not in history but in theoretical physics. It may seem decidedly odd that I have begun by recounting the beginning of the last century's major wars, and by trumpeting the capricious and convulsive character of human history. There is nothing new in the recognition that history follows tortuous paths, and that it has forever made a mockery of attempts to predict its course. My aim, however, is to convince you that we live in a special time, and that new ideas with a very unusual origin are beginning to make it possible to see *why* history is like it is; to see *why* it is and even must be punctuated by dramatic, unpredictable upheavals; and to see *why* all past efforts to perceive cycles, progressions, and understandable patterns of change in history have necessarily been doomed to failure.

A Faulty Peace

One may suspect that human history defies understanding because it depends on the unfathomable actions of human beings. Multiply individual unpredictability a billion times, and it is little wonder that there are no simple laws for history—nothing like Newton's laws, for instance, that might permit the historian to predict the course of the future. This conclusion seems plausible, and yet one should think carefully before leaping to it. If human history is subject to unpredictable upheavals, if its course is routinely and drastically altered by even the least significant of events, this does not make it unique as a process. In our world, these characteristics are ubiquitous, and it is just dawning on a few minds that there are very deep reasons for this.

The city of Kōbe is one of the gems of modern Japan. It lies along the southern edge of the largest Japanese island of Honshū, and from there its seaport—the world's sixth largest—handles each year nearly a third of all Japan's import and export trade. Kōbe has excellent schools, and its residents bask in what seems to be a haven of environmental stability. The city has good reason to call itself an "urban resort":[9] peaceful sunrises have for centuries given way to bright, warm afternoons, which have in turn slipped into cool, tranquil evenings. If visiting Kōbe, you would never guess that just beneath your feet invisible forces were preparing to unleash unimaginable violence. Unless, of course, you happened to be there at 5:45 A.M., January 17, 1995, when the calm suddenly fell to pieces.

At that moment, at a location just off the Japanese mainland, twenty kilometers southwest of Kōbe, a few small pieces of rock in the ocean floor suddenly crumbled. In itself, this was unremarkable; minor rearrangements of the Earth's crust happen every day in response to the stresses that build up slowly as continental plates, creeping over the planet's surface, rub against one another. But this time, what started as a minor rearrangement did not end up that way. The crumbling of those first few rocks altered the stresses on others nearby, causing them also to break apart. Farther down the line, still others followed suit, and in just fifteen seconds the earth ripped apart along a line some fifty kilometers long. The resulting earthquake shook the ground with the energy of a hundred nuclear bombs, ruining every major road or rail link near Kōbe and, in the city itself, causing more than a hundred thousand buildings to tilt or collapse. It sparked raging fires that took a week to control, and rendered inoperable all but 9 of the 186 berths in Kōbe's port. Ultimately, the devastation killed five thousand people, injured thirty thousand, and left three hundred thousand homeless.[10]

For centuries the area around Kōbe had been geologically quiet. Then, in just a few seconds, it exploded. Why?

Japan is known for its earthquakes. A quake releasing ten times as much energy leveled the city of Nobi in central Japan in 1891, and others struck in 1927, 1943, and 1948 at other locations. The intervals between these great earthquakes—thirty-five, sixteen, and five years—hardly form a simple, predictable sequence, as is typical of earthquakes everywhere. If the historian H. A. L. Fisher failed to see in history "a plot, a rhythm, a predetermined pattern," then so too have geophysicists failed utterly, despite immense effort, to discern any simple pattern in the Earth's seismic activity.

Modern scientists can chart the motions of distant comets or asteroids with stunning precision, yet something about the workings of the Earth makes predicting earthquakes extremely difficult, if not altogether impossible. Like the fabric of international politics, the Earth's crust is subject to sporadic and seemingly inexplicable cataclysms.

The Great Burnout

Not far to the west of Wyoming's vast Bighorn Basin, the wild and unrestrained landscape of Yellowstone National Park climbs into the Rockies. Immense forests of aspen and lodgepole pine clothe the mountains like a soft fabric, hiding black bears and grizzlies, moose, elk, deer, and innumerable species of birds and squirrels, all thriving in the seemingly pristine wilderness. Here and there a great rocky dome bursts out of the pines and towers over the park like a timeless sentinel. This is America's most beautiful natural park, set aside for protection back in 1872, and now the holiday destination of more than a million visitors each year.

But if Yellowstone is a place of almost unfathomable peace, it is also, sporadically, a place of terrific, incendiary violence.

Lightning sparks several hundred fires within the park every year. Most burn less than an acre, or maybe a few acres before dying out, while others carry on to destroy a few hundred or, far more rarely, a few thousand. As of 1988, even the largest fire ever recorded, in 1886, had burned only twenty-five thousand acres. So late in June of 1988, when a lightning bolt from a summer thunderstorm sparked a small fire near Yellowstone's southern boundary, no one was unduly alarmed. The fire was named the Shoshone, and the Forest Service began monitoring its progression. Within a week, storms had ignited a couple of other fires elsewhere in the park, and yet there was still no cause for concern. On July 10, when a brief rain fell, there were a handful of fires still smoldering, but all seemed well in hand and likely to burn out in the coming weeks. It didn't happen that way.

Whether it was the unusually dry conditions or the persistent winds, no one can really say, but by the middle of July the fires had only become bigger. "Up until then, with the fires," a National Park Service spokeswoman later recalled, "it was business as usual."[11] But on July 14, a fire given the name Clover spread to forty-seven hundred acres, and another called the Fan grew to cover twenty-nine hundred acres. Four days later yet another fire, sparked in an area known as Mink Creek, had exploded to cover thirteen thousand acres, and forest managers were beginning to see things that no expert had envisaged. The Shoshone fire suddenly gathered new life, racing to consume more than thirty thousand acres in just a few days, and by August some two hundred thousand acres of the park either had burned or were burning; on all fronts flames were advancing five to ten miles each day under a smothering blanket of smoke ten miles high.

Over the next two months, more than ten thousand firefighters from across the country, using 117 aircraft and more than a hundred fire engines, struggled ineffectually as the blaze swept through the park. Eventually the flames consumed 1.5 million acres and more than $120 million in federal firefighting money, and lost momentum and dwindled only with the coming of the first snow in autumn. Somehow, from one or several insignificant bolts of lightning an unstoppable inferno had emerged that made the previous worst fire in the history of Yellowstone look like a backyard barbecue. What made this one so bad? And why didn't anyone see it coming?

A Sharp Turn South

On September 23, 1987, investors around the world picked up *The Wall Street Journal* to see a headline heralding still more fantastic news: STOCK PRICES SOAR IN HEAVY TRADING; INDUSTRIALS RISE RECORD 75.23 POINTS.[12] It had been an incredible summer, and with almost perfect regularity, each day and each week had brought higher numbers and easy profits. Several weeks earlier, prices on the New York Stock Exchange had reached a new all-time high, and though they had slipped a tiny bit since then, the surge of September 23 was exactly what most traders expected. It was the natural end of a minor "correction," and it set the stage for further gains. "In a market like this," one trader said, "any news is good news. It's pretty much taken for granted now that the market is going to go up."[13]

So, two weeks later, when the market opened for trading on October 6, most analysts fully expected stock prices to climb even higher. When prices unexpectedly began to tumble, there was at first little concern. It was obvious to most analysts that this was merely another insignificant correction, a temporary setback caused by investors' uncertainty about interest rates or the value of the dollar. But for some reason this tiny correction took hold. By the end of the day, a sudden rash of selling had treated the market's optimistic "bulls" to a sharp rap on the nose. As one remarked,

> This one really came out of the blue. I didn't expect it to
> be so bad. . . . we froze around 3 P.M. and just started
> watching the screens. Even the phones stopped ringing.
> We were watching history in the making.[14]

Even so, the historical drama had really only begun.

Reassuringly, the press quickly pointed out that the drop of October 6, if considered in historical context, wasn't even one of the hundred largest in percentage terms. So it wasn't really all that serious. The market continued to slide over the next week, and then October 14, 15, and 16 saw three considerable losses in a row. Still, as traders left New York City for the weekend, they were reading a *Wall Street Journal* that remained stoutly confident and hopeful:

> It was the third major decline in as many days. But sev-
> eral technical analysts said that the big volume accompa-
> nying Friday's session might mean better things ahead.

In fact, it meant something rather more ominous.

Over that weekend, in the minds and hearts of thousands of major investors, the subtle apprehensions that had been accumulating over the previous weeks precipitated into a torrential rain of fear. On Black Monday, October 19, when the exchange opened for trading at 9:30 A.M., it was immediately swept away in a mad panic: prices began to plummet. The rush to sell was so overwhelming that, by late afternoon, the outstanding value of stocks and bonds had been clipped by more than one-fifth, and some $500 billion had been erased from investors' financial statements. At close, a nightmarish gloom settled over Wall Street as traders contemplated the largest single-day free fall in market history. "It felt like the end of the world," wrote *Newsweek*, "after two generations of assurances that it couldn't possibly happen." The crash was nearly twice as severe as the infamous stock market collapse of 1929, although this time, fortunately, it didn't trigger a global economic depression. "It was God tapping us on the shoulder," one billionaire investor concluded, "a warning to get our act together."

As with the First World War or the great quake in Kōbe, no one had predicted the crash. Immediately *afterward*, on the other hand, analysts produced all kinds of dubious explanations for why it had happened when it did. Even today, however, there is little consensus. As a longtime Wall Street analyst has recently concluded,

> The crash of 1987 was such a storm of mass emotion that "market as machine" theorists worked overtime explaining the drop and figuring out how to "fix" the system. The theory that gained the most credence was that the crash was caused by so-called portfolio insurance computer programs, which in essence sold stocks as the market went lower. . . . Unfortunately for the theory, it does not explain very well why markets around the world crashed simultaneously or why the decline stopped. It is at an utter loss to explain why many indexes around the world that had no computer trading fell further than the Dow Jones Industrial Index. It also ignores the fact that

throughout 1986 and 1987, market observers in an equally serious tone had continually explained why a stock market crash was impossible because of "safeguards that are in place," safeguards such as portfolio insurance.[15]

On a Sharp Edge

The roots of war are to be sought in politics and history, those of earthquakes in geophysics, of forest fires in patterns of weather and in the natural ecology, and those of market crashes in the principles of finance, economics, and the psychology of human behavior. Beyond the labels "disaster" and "upheaval," each of these events erupted from the soil of its own peculiar setting. Still, there is an intriguing similarity. In each case, it seems, the organization of the system—the web of international relations, the fabric of the forests or of the Earth's crust, or the network of linked expectations and trading perspectives of investors—made it possible for a small shock to trigger a response out of all proportion to itself. It is as if these systems had been poised on some knife-edge of instability, merely waiting to be "set off."

In the history of life, we find a similar pattern. The fossil record reveals that the number of species on our planet has—roughly speaking—grown steadily over the past six hundred million years. Yet on at least five separate occasions, sudden and terrible mass extinctions nearly wiped out every living thing. What happened? Many scientists point to precipitous changes in the Earth's climate, caused perhaps by the impact of large asteroids or comets. Others suggest that the extinction of just a single species can, on occasion, trigger others, which in turn cause still others, leading to an avalanche of extinctions that can consume large fractions of entire ecosystems. The mass extinctions continue to mystify biologists and geologists, and yet one thing is clear: if the fabric of life seems resilient and largely in balance with itself, the truth is rather more unsettling. The global ecosystem is occasionally visited by abrupt episodes of collapse.

When I was in grade school, one of the dreaded tasks assigned by the geometry teacher was to determine if two triangles were *similar*. Here is a big triangle, she would say, and here is another much smaller triangle, oriented in a different way. Are they, aside from the irrelevant details of overall size and orientation, the same triangle? Put otherwise: If you can shrink or expand either triangle at will, turn the two over and rotate them in any way you like, can you make the one fit precisely over the other? If so, then the triangles are similar—if you understand the essential logic of one, its angles and the ratios of the lengths of its sides, then you also understand the other.

Three centuries ago, Isaac Newton sparked a scientific revolution by noticing another kind of similarity. His contemporaries must have been at first disbelieving, and later stunned, when he told them that an apple falls to the ground in precisely the same way as the Earth moves round the Sun. Newton saw that both Earth and apple fall into the single category of *things moving under the force of gravity*. Before Newton, happenings on Earth and in the Heavens were utterly incomparable. Afterward, the motions of an apple or an arrow, a satellite, or even an entire galaxy were seen as deeply similar—as mere instances of a single, deeper process.

"The art of being wise," the American philosopher and psychologist William James once wrote, "is the art of knowing what to overlook,"[16] and this book is about a terrific step along the scientific road of learning what to overlook. It is about the discovery of a profound similarity not between triangles or moving objects, but between the upheavals that affect our lives, and the ways in which the complicated networks in which they occur—economies, political systems, ecosystems, and so on—are naturally organized. We might add to our list dramatic changes in fashion or musical taste, episodes of social unrest, technological change, even great scientific revolutions.

As we shall see, the key to a unified understanding lies in the subtle and powerful concept of the *critical state*, an idea that appears to be central to the scientific understanding of many processes in which the notion of history plays a fundamental role.

History Matters

For centuries, physicists have sought to capture the fundamental laws of the universe in timeless and unchanging equations, such as those of quantum theory or relativity. While this project has been enormously successful, the ultimate simplicity of such equations points to a paradox: If the laws of physics are so simple, why is the world so complex? Why don't ecosystems, organisms, and economies reveal the same simplicity as Newton's laws and the other laws of physics?

In the late 1970s and 1980s, scientists discovered at least part of the answer—*chaos*. When a pinball scatters through a pinball machine, its path is extraordinarily sensitive to tiny influences along the way. This is chaos. Inside any ordinary balloon, the molecules also move according to the law of chaos: give a tiny nudge to just a single molecule, and in much less than a minute every last one will be affected. In the context of the Earth's atmosphere, chaos brings us the "butterfly effect," the incredible conclusion that the flapping of a butterfly's wings in Portugal now might just lead to the formation of a severe thunderstorm over Moscow in a couple of weeks' time.

So here we have one mechanism by which complexity can grow out of simplicity. Predicting the long-term future of any chaotic system is practically impossible, and a chaotic process looks wildly erratic even if the underlying rules are actually quite simple. Researchers have discovered chaos at work in the fluctuations of things ranging from lasers to rabbit populations, and in the late 1980s and early 1990s some scientists even hoped that chaos might finally make sense of the wild ups and downs of financial markets.

But it didn't, for there is an aspect of the world's complexity that chaos leaves completely untouched. Not its unpredictability, but its *upheavability*. Chaos is limited in its ability to explain tumultuous events, as researchers had hoped it might, because chaos in itself does not generate upheavals. Something more is needed if chaos is to give rise to tumultuous events, such as stock market crashes or earthquakes, and that something more is history.

To see why, think again about the famous butterfly example, but with one important difference: imagine that the butterfly is inside a balloon. A butterfly could flap its wings for eternity inside a balloon and never cause the equivalent of a thunderstorm in that enclosed space. This is because the air in the balloon lives in peace under unchanging conditions, in what scientists refer to as *equilibrium*. In equilibrium, it is certainly true that the individual molecules toss around in utter chaos, but that's pretty much the end of the story. No larger patterns ever emerge, nothing important ever happens, and so the idea of history has little meaning. For the air in the balloon, the past and the future are essentially the same. In contrast, the air in the Earth's atmosphere is very much out of equilibrium. Far from being left in peace, it is being continually stirred and agitated and energized by the influx of light from the Sun. The result is the rich and ever unfolding history of the weather and climate. Out of equilibrium, there is such a thing as history.

This gives us a clue about the cause of upheavability: it clearly has something to do with the way things work when out of equilibrium. For the most part, out-of-equilibrium physics remains a forest of the unknown. And yet, over the past two decades, scientists have forged a few remarkable insights, one of which casts the upheavals we met earlier in this chapter in a fascinating light. The key idea is the notion of the *critical state*, a special kind of organization characterized by a tendency toward sudden and tumultuous changes, an organization that seems to arise naturally under diverse conditions when a system gets pushed away from equilibrium. This is the first landmark discovery in the emerging science of nonequilibrium physics—what we might equally well call the field of *historical physics*.

I should point out that in recent years this field has also gone under another name: complexity theory. After all, when things are out of equilibrium they tend to be complex—the intricately knitted structure of a food web, the irregular surface of a fractured brick, the infinitely detailed shape of a snowflake. But *history* reveals the essential element that underpins complexity in all these cases. You cannot understand either a food web or a snowflake by solving simple, timeless equations. Instead, you need to delve into the past and deal with a long, tortuous history carved out by what the biologist Francis Crick has termed "frozen accidents." If some species goes extinct, this leaves an unalterable mark on the food web forevermore. If at some instant a small bit of water freezes on one side of a growing snowflake rather than another, its future is likewise altered irreversibly. Such accidents, irreversible in their consequences and piling up one on the other, lead to the complexity we see around us.

But if complexity emerges out of strings of historical accidents, and there are no fundamental equations for things in which history matters, how can one achieve any scientific understanding? Scientists have recently found a way—by replacing equations with games. The physics research journals are now stuffed with papers about the workings of simple mathematical games: some meant to explore the basic historical process behind crystal growth, others to mimic that which lies behind the formation of rough surfaces, and so on. There are hundreds, each slightly different in its details, but all sharing a deeply historical nature. These games offer a way to proceed in the face of history and its messy strings of accidents. In effect, they permit scientists to greatly simplify the things they're studying, whether an economy or an ecosystem, and to focus on the fundamental processes at work without being distracted by myriad confusing details.

And of all these games, one stands out as a kind of archetype of simplicity, and has been central to the discovery of the underlying cause of a vast range of tumultuous events. To understand this "sand-pile game," a focal point for our story, imagine dropping grains of sand one by one onto a table and watching the pile grow. A grain falls accidentally here or there, and then in time the pile grows over it, freezing it in place. Afterward, the pile feels forever more the influence of that grain being just where it is and not elsewhere. In this case, clearly, history matters, since what happens now can never be washed away, but affects the entire course of the future.

"All great deeds and all great thoughts," Albert Camus once wrote, "have ridiculous beginnings."[17] And so it was in 1987 when physicists Per Bak, Chao Tang, and Kurt Weisenfeld began playing this sandpile game in an office at Brookhaven National Laboratory, in New York State. As it turns out, this seemingly trivial game lies behind the discovery of the widespread importance of the critical state—the discovery that can help us to make sense of upheavals.

The Sand Men and the Critical State

Theoretical physicists enjoy posing seemingly trivial questions that, after a bit of thinking, turn out not to be so trivial. In this respect the sandpile game turned out to be a real winner. As grains pile up, it seems clear that a broad mountain of sand should edge slowly sky-ward, and yet things obviously cannot continue in this way. As the pile grows its sides become steeper, and it becomes more likely that the next falling grain will trigger an avalanche. Sand would then slide downhill to some flatter region below, making the mountain smaller, not bigger. As a result, the mountain should alternately grow and shrink, its jagged silhouette forever fluctuating.

Bak, Tang, and Weisenfeld wanted to understand those fluctuations: What is the typical rhythm of the growing and shrinking sand-pile? Of course, they didn't really care about sandpiles. In studying this silly problem, they were really chasing some insights regarding the general workings of nonequilibrium systems. The sandpile seemed like a nice, simple starting point, and with luck, they hoped, they might discover in this setting some patterns of behavior that would apply to a lot more than just sandpiles.

Unfortunately, dropping sand one grain at a time is a delicate and laborious business. So in seeking some answers concerning the rhythm of the pile's growth, Bak and his colleagues turned to the computer. They instructed it to drop imaginary "grains" onto an imaginary "table," with simple rules dictating how grains would topple downhill as the pile grew steeper. It was not quite the same as a real sandpile, and yet the computer had one spectacular advantage: a pile would grow in seconds rather than days. It was so easy to play the game that the three physicists soon became glued to their computer screens, obsessed with dropping grains and watching the results. And they began to see some curious things.

The first big surprise came as the answer to a simple question: What is the typical size of an avalanche? How big, that is, should you expect the very next avalanche to be? The researchers ran a huge number of tests, counting the grains in millions of avalanches in thousands of sandpiles, looking for the typical number involved. The result? Well . . . there was no result, for there simply was no "typical" avalanche. Some involved a single grain; others ten, a hundred, or a thousand. Still others were pile-wide cataclysms involving millions that brought nearly the whole mountain tumbling down. At any time, literally anything, it seemed, might be just about to happen.

Imagine wandering into the street, anticipating how tall the next person might be. If people's heights worked like these avalanches, then the next person might be less than an inch tall, or over a mile high. You might crush the next person like an insect before seeing him or her. Or imagine that the duration of your trips home from work went this way; you'd be unable to plan your life, since tomorrow evening's journey might take anything from a few seconds to a few years. This is a rather dramatic kind of unpredictability, to say the least.

To find out why it should show up in their sandpile game, Bak and colleagues next played a trick with their computer. Imagine peering down on the pile from above, and coloring it in according to its steepness. Where it is relatively flat and stable, color it green; where steep and, in avalanche terms, "ready to go," color it red. What do you see? They found that at the outset the pile looked mostly green, but that, as the pile grew, the green became infiltrated with ever more red. With more grains, the scattering of red danger spots grew until a dense skeleton of instability ran through the pile. Here then was a clue to its peculiar behavior: a grain falling on a red spot can, by dominolike action, cause sliding at other nearby red spots. If the red network was sparse, and all trouble spots were well isolated one from the other, then a single grain could have only limited repercussions. But when the red spots come to riddle the pile, the consequences of the next grain become fiendishly unpredictable. It might trigger only a few tumblings, or it might instead set off a cataclysmic chain reaction involving millions. The sandpile seemed to have configured itself into a hypersensitive and peculiarly unstable condition in which the next falling grain could trigger a response of any size whatsoever.

This may seem like something that only a physicist could find interesting. After all, in other settings, scientists have known about this condition for more than a century; they have referred to it technically as a critical state. But to physicists, it has always been seen as a kind of theoretical freak and sideshow, a devilishly unstable and unusual condition that arises only under the most exceptional circumstances—in liquids, for example, when held at precise temperatures and pressures under extraordinarily well controlled laboratory conditions. In the sandpile game, however, a critical state seemed to arise naturally and inevitably through the mindless sprinkling of grains.[18]

This led Bak, Tang, and Weisenfeld to ponder a provocative possibility: If the critical state arises so easily and inevitably in a simple computer model of a growing sandpile, might something like it also arise elsewhere? Despite what scientists had previously believed, might the critical state in fact be quite common? Could riddling lines of instability of a logically equivalent sort run through the Earth's crust, for example, through forests and ecosystems, and perhaps even through the somewhat more abstract "fabric" of our economies? Think of those first few crumbling rocks near Kōbe, or that first insignificant dip in prices that triggered the stock market crash of 1987. Might these have been "sand grains" acting at another level? Could the special organization of the critical state explain why the world at large seems so susceptible to unpredictable upheavals?

A decade of research by hundreds of other physicists has explored this question and taken the initial idea much further.[19] There are many subtleties and twists in the story to which we shall come later in this book, but the basic message, roughly speaking, is simple: The peculiar and exceptionally unstable organization of the critical state does indeed seem to be ubiquitous in our world. Researchers in the past few years have found its mathematical fingerprints in the workings of all the upheavals I've mentioned so far, as well as in the spreading of epidemics, the flaring of traffic jams, the patterns by which instructions trickle down from managers to workers in an office, and in many other things.[20] At the heart of our story, then, lies the discovery that networks of things of all kinds—atoms, molecules, species, people, and even ideas—have a marked tendency to organize themselves along similar lines. On the basis of this insight, scientists are finally beginning to fathom what lies behind tumultuous events of all sorts, and to see patterns at work where they have never seen them before.

Critical World?

So the ubiquity of the critical state might well be considered the first really solid discovery of complexity theory—or of what I have been calling historical physics. This is a discovery with implications, and not only for physicists and other scientists.[21] If the laws of physics didn't allow "frozen" accidents, the world would be in equilibrium, and everything would be like the gas in a balloon, resting forever in the same uniform and unchanging condition. But the laws of physics do allow events to have consequences that can become locked in place, and so alter the playing field on which the future unfolds. The laws of physics allow history to exist, and to play a crucial role in the way our world works. The discovery of the ubiquity of the critical state, then, is also the first deep discovery concerning the way that historical processes usually work, which brings us back to the point from which we started this chapter.

In principle, history could unfold far more predictably than it does. It needn't, in principle, be subject to terrific cataclysms of all sorts. One of our tasks in this book is to examine why the character of human history is as it is, and not otherwise. The answer, I suggest, is to be found in the critical state and in the new nonequilibrium science of games, which aims to study and categorize the kinds of historical processes that are *possible*. If many historians have searched for gradual trends or cycles as a way of finding meaning and making sense of history, then they were using the wrong tools. These notions arise in equilibrium physics and astronomy. The proper tools are to be found in nonequilibrium physics, which is specifically tuned to understanding things in which history matters.

In the very same year that Bak, Tang, and Weisenfeld invented their game, the historian Paul Kennedy published *The Rise and Fall of the Great Powers*.[22] In that book he laid out the idea that the large-scale historical rhythm of our world is determined by the natural buildup and release of stress in the global network of politics and economics. His view of the dynamics of history leaves little room for the influence of "great individuals," and is more in keeping with the words of John Kenneth Galbraith quoted at the beginning of this chapter. It sees individuals as products of their time, having limited freedom to respond in the face of powerful forces. Kennedy's thesis, in essence, is this: The economic power of a nation naturally waxes and wanes. As times change, some nations are left clinging to power that their economic base can no longer support; others find new economic strength, and naturally seek greater influence. The inevitable result? Tension, which grows until something gives way. Usually the stress finds its release through armed conflict, after which the influence of each nation is brought back into rough balance with its true economic strength.

If this sounds at all like the processes at work in the Earth's crust, where stresses build up slowly to be released in sudden earthquakes, or in the sandpile game, where the slopes grow higher and more unstable until leveled again in some avalanche, it may be no coincidence. We shall see later that wars actually occur with the same statistical pattern as do earthquakes or avalanches in the sandpile game. Kennedy could find strong support for his thesis—as well as a more adequate language in which to describe it—in this theoretical idea. He may have been struggling to express in words, and in a historical context, what the concept of the critical state expresses mathematically.

Whatever lessons historians may be able to draw from all this, the meaning for the individual is more ambiguous. For if the world is organized into a critical state, or something much like it, then even the smallest forces can have tremendous effects. In our social and cultural networks, there can be no isolated act, for our world is designed—not by us, but by the forces of nature—so that even the tiniest of acts will be amplified and registered by the larger world. The individual, then, has power, and yet the nature of that power reflects a kind of irreducible existential predicament. If every individual act may ultimately have great consequences, those consequences are almost entirely unforeseeable. Out there right now on some red square in the field of history a grain may be about to fall. Someone trying to bring warring parties to terms may succeed, or may instead spark a conflagration. Someone trying to stir up conflict may usher in a lengthy term of peace. In our world, beginnings bear little relationship to endings, and Albert Camus was right: "All great deeds and all great thoughts have ridiculous beginnings."

One of the inevitable themes of our story is that if one wants to learn about the rhythms of history (or, shall we say, its disrhythms), one might just as well become familiar with the process by which, say, earthquakes happen. If the organization of upheaval and hypersensitivity is everywhere, one need not look far to find it. So let us leave human history and the individual aside for the moment, and first look to the simpler world of inanimate things. Let us go underground, into the dark, gritty world beneath the Earth's surface, and take a closer look at what goes on there. Surprisingly, in the underworld rumblings of our changeable planet, we shall encounter a way to understand the workings of a thousand things.

· 2 ·

A Shaky Game

"Science" means simply the aggregate of all the recipes that are always successful. All the rest is literature.

—PAUL VALÉRY[1]

———————

Since my first attachment to seismology, I have had a horror of [earthquake] predictions and predictors. Journalists and the general public rush to any suggestion of earthquake prediction like hogs toward a full trough.

—CHARLES RICHTER[2]

♦

IN APPROACHING THE PAST, DIFFERENT HISTORIANS AIM TO DO DIFFERENT things; indeed, for centuries historians have been arguing among themselves about what they ought to be doing. But certainly one aim of many historians has been to trace the general causes and conditions that have made important human events—wars, revolutions, and so on—take place at particular times and in particular settings. One of our deepest beliefs about the world is that great events do not rise up "out of nowhere." If there is a great explosion in the center of a city, one assumes that there was a great explosive planted there beforehand. We all share the "common sense" habit of mind that longs to discover significant causes behind great events of all kinds.

Likewise, when a terrific earthquake strikes, or when a volcano suddenly erupts, geophysicists want answers. What made it happen? Were there details in the earth that could have warned us? For some volcanic eruptions, at least, this seems to be the case. The violent eruption of Mount St. Helens in Washington State in 1980 was preceded by "visible ground deformation of up to one meter per day, by eruptions of gas and steam, and by thousands of small earthquakes,"[3] all leading to the devastating explosion that blew apart the mountain. Because of these precursors, officials were able to warn the public a month before it happened.

For more than a century, scientists have been searching for similar signatures of the special conditions that hold just before large earthquakes. But in this search, however, they have faced perpetual frustration.

Mission Impossible?

Anyone would have been dumbfounded to walk through the streets of downtown St. Louis in the waning days of November 1990. Here it was, less than one month before Christmas, and by rights the shops should have been crammed with people buying gifts and collecting decorations for Christmas trees. But in the winter of 1990, the big department stores were barren and the city streets were virtually empty. The action was elsewhere, in the supermarkets and hardware stores of the suburbs, where millions of people were buying drinking water and canned goods rather than gifts, and stocking up on candles, flashlights, blankets, shovels, and electrical generators. It was all over the newspapers, and seemed a near certainty: sometime between December 1 and December 5, St. Louis was going to be rocked by a tremendous earthquake.

The people of St. Louis weren't the only ones in near hysteria. All across the mideastern United States that winter, in Illinois, Arkansas, Tennessee, and elsewhere, people were scrambling in fear and excitement to get ready. Local and state authorities established plans for handling the impending catastrophe. Schools were to be closed in several states, and emergency crews stood on high alert. Armies of volunteers were organized and given specific responsibilities: to deliver water, to set up makeshift hospitals, to help firefighters. In St. Louis alone, officials expected at least three hundred deaths, and damage to buildings to exceed $600 million.

The earthquake had been predicted by a "business consultant and climatologist" named Iben Browning. As it happened, Browning was a scientist with a Ph.D., and perhaps this explains why the media took his prediction so seriously—even though his Ph.D. was in biology. Browning claimed that on the dates in question, the Sun, Earth, and Moon would be in rare alignment, and that their combined gravity would create tidal forces that would stress the rocks in the New Madrid fault zone past their limit. In Missouri, the state spent

$200,000 getting ready. In St. Louis, homeowners coughed up an additional $22 million to beef up their insurance coverage.

All through the Browning affair, responsible scientists from the U.S. Geological Survey (U.S.G.S.) and from universities in the area insisted that the prediction had no scientific merit. One report criticized Browning for making

> a leap from hypothesis to prediction without the intervening process of showing verifiable evidence and hypothesis testing that makes mainstream science a viable and successful discipline.[4]

And sure enough, Browning's Great Earthquake of 1990 never happened.

Unlike Browning, mainstream geophysicists are cautious to the point of paranoia when it comes to making any earthquake prediction. This is not just due to their healthy respect for the social disruption that any prediction causes. In truth, it is the direct result of a very long string of embarrassing failures in earthquake prediction by geophysicists themselves. After a century of research, virtually all quakes still come completely unannounced. No one predicted the 1995 earthquake in Kōbe, even though Japan has a long-standing and well-funded program in earthquake prediction. More embarrassing yet are the quakes that scientists have predicted, and that, like Browning's, haven't shown up.

Uncertain Ground

In the late 1970s, Japanese scientists were sure that a "great Tokai earthquake" was soon to hit central Japan. As one researcher put it,

> Many Japanese seismologists, earthquake engineers, and national and local officials responsible for disaster pre-

vention are quite convinced nowadays that a great quake
of magnitude 8 or so will hit the Tokai area . . . in central
Japan between Tokyo and Nagoya in the near future. . . .
The targeted area was often struck by great earthquakes
in historical times such as the 1854 and 1707 earthquakes. . . . The mean return period of recurrence of
great earthquakes there is estimated at about 120 years.
As more than 120 years have already passed since the last
shock, there is reason to believe an earthquake will recur
sooner or later.[5]

The reasoning here is simple. There must, it is thought, be a "typical" time between earthquakes. If in some region more than the typical time has passed since the last earthquake, the next must be
overdue. Believing this idea, Japanese authorities in the 1970s set up
an early warning system. Any odd seismic data would trigger an
emergency meeting of an "earthquake assessment committee" that
would decide whether to shut down nuclear reactors, expressways,
schools, and factories. Ever since, the Japanese have practiced
responding to the alarm, one day each year on the anniversary of the
great Kanto earthquake of 1923. But decades later, there has been no
Tokai earthquake. Not even a murmur. The Kōbe quake occurred in
an area where the authorities thought the risk was small.

In 1976 Brian Brady at the U.S. Bureau of Mines predicted that
two enormous quakes of magnitude 9.8 and 8.8 would strike off the
coast of Peru in August 1981 and May 1982. He also claimed that a
large foreshock of magnitude 7.5 to 8 would precede these events in
June 1981.[6] A humiliated Brady retracted his prediction when the
foreshock didn't happen. But the Peruvian government was already
thrown into such a scare that an official of the U.S. Geographical
Survey had to travel there to calm their fears.

In addition to searching for precursors, scientists have tried to
identify reliable cycles in earthquake activity. The Moon, the Sun,

the planets, the days—we live in a world of repetitive cycles, many of astronomical origin. Perhaps some earthquakes work this way too? If so, then predicting—at least at some places on Earth—would be easy. Earth scientists latched on to this seemingly plausible idea in the mid-1980s, and so unwittingly set themselves up for one of their most demoralizing failures.

The San Andreas Fault runs down the western edge of California. Along the way, it is plainly visible from the air as a peculiar straight line running north and south through the hills. The ground along the western side of the fault is creeping slowly northward— two to three centimeters each year—relative to the ground on the eastern side. The state of California, therefore, isn't just one monolithic slab of earth, but two, pushing slowly past one another. This motion isn't usually apparent, since the two slabs tend to stick together by friction, much as a piece of furniture sticks to the floor when you try to slide it across. Push the furniture hard enough, however, and it will eventually slide—usually all at once in a sudden burst. So it goes with the two pieces of the Earth's crust along the San Andreas Fault. They usually stick, but occasionally slip, and this triggers an earthquake.

In 1979, geophysicist William Bakun and some of his colleagues at the U.S. Geological Survey in Menlo Park, California, noticed something interesting about the record of past earthquakes on a small segment of the San Andreas Fault near the rural community of Parkfield, about 150 miles south of San Francisco. One quake happened there in 1966. Another happened in 1934, and, going back further, there were others in 1922, 1901, 1881, and 1857. Counting the numbers of years between these quakes, the U.S.G.S. researchers found the fairly regular sequence: 24, 20, 21, 12, and 32. Not only did numbers close to 20 appear frequently, but the average time between quakes was twenty-two years. There was something else intriguing about these earthquakes, too. Geophysicists measure the size of an earthquake by its magnitude, a number, such as 5 or 7, or

6.4, that reflects how violently the ground shakes near the quake, or, equivalently, how much energy it releases.[7] Bakun and his colleagues noticed that all the Parkfield quakes had magnitudes between 5.5 and 6. The implication seemed clear: these earthquakes must arise from some cyclic, repetitive process. Perhaps, they reasoned, it takes about twenty-two years for the stress to build to the breaking point, and then the rocks along this section of the fault slip.

Geophysicists the world over were soon convinced that the fault near Parkfield was the earthquake equivalent of Old Faithful. Like that famous geyser in Yellowstone that erupts every hour or so, the Parkfield earthquake seemed to strike every 22 years or so. Since the last quake hit in 1966, the next should have been due in 1988. The field of earthquake prediction finally seemed on the verge of a long-overdue breakthrough.[8]

After an international panel of experts judged the prediction of Bakun and his colleagues sound, the director of the U.S. Geological Survey issued a rare public prediction on April 5, 1985, saying that a quake should occur near Parkfield within the next five to six years.[9] Researchers strapped the hills of the region with the densest and most sophisticated array of earthquake monitoring devices anywhere in the world, and began waiting. In 1986, the Board on Earth Sciences of the U.S. National Research Council summed up scientists' confidence:

> Nowhere else in the world is a prediction in effect with a degree of confidence as high as that for Parkfield. Here, on a specific 25-kilometer segment of the San Andreas fault . . . studies during the past decade indicate a 95 percent probability that an earthquake of about magnitude 6 will occur between 1986 and 1993.[10]

The stage was set. A 1987 article in *The Economist* referred to Parkfield as "geophysicists' Waterloo." If the earthquake didn't arrive as expected, "then earthquakes are unpredictable and science is defeated. There will be no excuses left, for never has an ambush been more carefully laid for such an event."[11]

To geophysicists' dismay, the earthquake didn't arrive. Despite that impressively regular string of past earthquakes at Parkfield, no earthquake of magnitude 5.5–6 has struck the Parkfield area to this day. If one arrived now, it would be too late. The prediction would still be wrong. It seems that researchers were duped by their very human desire to discover regular cycles where there were none.

As geophysicist Yakov Kagan of the University of California at Los Angeles has pointed out,[12] there is a wealth of places on Earth where earthquakes happen, and where geophysicists study them. Suppose that the earthquakes strike randomly, with no intelligible pattern whatsoever. Even then you would expect once in a while for a string of quakes to hit somewhere in a highly regular sequence, just by pure chance. And since there are so many places on Earth where quakes occur, a diligent search through the list is likely to turn up a regular sequence somewhere. That somewhere happened to be Parkfield.

All these predictions, made by scientists working in the mainstream of geoscience, might remind us of Mark Twain's remark, "There is something fascinating about science. One gets such wholesale returns of conjecture out of such a trifling investment of fact."[13] In 1997, when geophysicist and earthquake expert Robert Geller of the University of Tokyo wrote a long review article summarizing the state of earthquake prediction, he referred to more than seven hundred research papers written over the past half century, many claiming to have identified some kind of precursor. Geller concluded that not a single one is even close to being reliable. An earthquake prediction worth its salt would foretell the location, timing, and size of the quake. And since closing schools and factories and evacuating cities is a costly business, it has been estimated that a useful prediction "would need a 50% chance of being right, and an accuracy of one day in time, and about 50 kilometers in space."[14] Researchers would also need to get the severity of the earthquake reasonably correct. Something like this would be an enormous benefit to the people in the region affected.[15]

How close are we to being able to do this? Scientists nowadays have fabulous technology that permits measuring the rising, falling, or shifting of portions of the Earth's crust with centimeter accuracy. Yet in his 1997 review, Geller could only be pessimistic:

> Earthquake prediction research has been conducted for over 100 years with no obvious successes. Claims of breakthroughs have failed to withstand scrutiny. Extensive searches have failed to find reliable precursors . . . reliable issuing of alarms of imminent large earthquakes appears to be effectively impossible.[16]

If science is "the aggregate of all the recipes that are always successful," and the rest is literature, then we must conclude that there is no science of earthquakes, since there are no successful recipes. When it comes to predicting earthquakes, there is only literature. A full century of research has apparently amounted to nothing.

The Great Mud Ball

So what historians say about human history seems to apply to the workings of the Earth's crust as well. Major wars or revolutions don't come in simple cycles, and they don't telegraph their punches. The conditions that precede them are always different, and no one has ever identified any reliable precursors. As one historian has written,

> Time and again, history has proved a very bad predictor of future events. This is because history never repeats itself; nothing in human society . . . ever happens twice

under exactly the same conditions or in exactly the same way.[17]

So it is also with earthquakes. There are no cycles, no warnings, no signals, no precursors. The Earth starts shaking whenever it wants to.

Scientists can predict where and when a hurricane will hit, and how destructive it will be—at least roughly. We know when the atmospheric conditions are right for the formation of tornadoes, and when flooding is imminent on large rivers. In each case it is simply a matter of tracing the relevant precursors—storm clouds, winds, and rains—and sounding the alarm bells when things look ominous. Meteorologists even have a fair degree of success in predicting the weather on a daily basis, at least a few days in advance. But with earthquakes, similar understanding seems almost beyond science. Why?

To find some clues, we need to look in a bit more detail at what is going on inside the Earth, and at the natural buildup and release of stresses and strains in its crust. Geophysicists are particularly frustrated by their lack of prediction success, since they do know a great deal about the mechanics of earthquakes. Historians can be baffled with a clean conscience, as they are dealing with the deep mystery of human behavior. But the process that drives earthquakes is no mystery. Everything going on in the ground is fully deterministic, and fully predictable in principle. There is not much more to it than rock pressing against rock.

To form a picture of what our planet is like, inside and out, imagine a giant ball of wet mud set out to dry in the warm air. After a while, the outside will dry, forming a crusty shell, while on the inside the mud remains liquid.[18] This is rather like the Earth with one difference: The liquid inside the Earth is moving, whereas that inside the mud ball is stationary. If by some clever means you could stir up the liquid mud in the middle so that it would flow about beneath the surface, you would have a crude model for the Earth. Our planet has an outer shell of hard crusty material—the Earth's crust—covering hot material flowing about beneath—the mantle.

The motion is driven by tremendous heat in the Earth's interior. Naturally, the warmer material tends to rise and the cooler to sink. As a result, the Earth's crust floats on a deep ocean of mantle containing vast undercurrents. Being solid, the crust cannot move about continuously like the hot flowing mantle beneath. Instead it has broken into huge fragments, which glide over the Earth like gigantic rafts. In the modern theory of this movement—known as the theory of plate tectonics—the word "plate" refers to these fragmented pieces of the Earth's brittle crust, each some one hundred kilometers thick.

In some places the plates collide head-on with one another. This is happening beneath Japan, for example, where three separate plates are smashing together. In other places two plates rub shoulders along a long boundary, as in California. Here the Pacific Plate underlies most of the Pacific Ocean, as well as the tiny sliver of California to the west of the San Andreas Fault. To the east of the San Andreas, the crust belongs to the North American Plate, which underlies all of North America and the western half of the Atlantic Ocean. The North American Plate is slowly moving roughly southward. Meanwhile, the Pacific Plate is on its way to the north. The result is a terrific rubbing together all along the fault.

As we have already seen, the basic mechanism of earthquakes is just the sticking and slipping of one piece of rock as it tries to slide past another. The inexorable movement of the plates cannot be resisted forever. When rocks get so twisted out of shape, when the stress builds up past some crucial threshold, then surfaces suddenly slip, and there is an earthquake. It's been happening on the San Andreas Fault for fifteen to twenty million years, and it happens everywhere else where two or more plates come into contact. If you take a globe and draw a black dot to mark the location of every large quake over the past century, you find a remarkable pattern. Swarms of black dots appear at all of the boundaries between plates. Indeed, as you draw the dots more densely they form an image of the fractured surface of the planet, bringing the separate plates into rough relief.

There are, of course, many complications to this simple picture. To begin with, there are other kinds of collisions between plates apart from the two I have just mentioned. In some places plates are moving away from one another to make way for new hot material issuing up to the crust from below, or one plate is sliding beneath another, forcing old crust to be recycled and sent back into the melting pot of the mantle. There are, in all, eight major plates on the Earth, plus a variety of smaller ones, all being pushed about by the hidden conveyor action of the mantle currents below. What's more, while the earthquake-generating process is just sticking and slipping, the overall structure of the Earth's crust is extremely complicated, and not only because there are so many plates.

There are also hundreds of kinds of rock making up each plate, and they all have different properties. In some places the rock is strong; in others it isn't. It is safe to say that no two earthquake zones in the world are quite alike. Things are complicated further because all of the many faults that run over the crust interact with one another. Slippage along one fault will affect what happens on others.

Once we appreciate all of this complexity, we may no longer be surprised that accurate predictions cannot be made. And the story about earthquakes is actually even murkier than this.

Background Noise

On December 16, 1811, the first in a rapid succession of three terrific earthquakes hit the region surrounding New Madrid, Missouri, about 150 miles southeast of St. Louis. The first was so powerful that it rang church bells 1,000 miles away in Boston, altered the geography of a large portion of Missouri and Tennessee, and made the great Mississippi River run backwards. An eyewitness reported that

> . . . a very awful noise resembling loud but distant thunder, but more hoarse and vibrating . . . was followed in a few minutes by the complete saturation of the atmosphere with sulphurious vapor. . . . The screams of the affrighted inhabitants running to and fro, not knowing where to go, or what to do—the cries of the fowls and beasts of every species—the cracking of trees falling, and the roaring of the Mississippi—the current of which was retrograde for a few minutes . . . formed a scene truly horrible. . . .[19]

This earthquake and two subsequent others—on January 23 and February 7, 1812—were together powerful enough to dig out a new lake. Reelfoot Lake, located in Tennessee about 100 miles northeast of Memphis, did not exist in 1810.

If asked to imagine an earthquake, most people will conjure up images of something like these New Madrid quakes: cataclysms of almost unimaginable violence. And yet most earthquakes are not even remotely like this. If you visit the website of the U.S.G.S., you will find an up-to-date record of all the earthquakes that have struck along the San Andreas fault in northern California in the past days and weeks, and even in the past hour. The organization's seismographs send data to this site almost as quickly as it is recorded. You will not have heard about these quakes on the news, however, for almost all are less than magnitude 3. Remember, a magnitude 3 earthquake is ten thousand times less violent than a magnitude 7. It wouldn't knock a fly off the window. On August 30, 1999, for example, a typical day, no fewer than twenty-two earthquakes struck in California.[20] Only one reached magnitude 3. In each of these tiny earthquakes, rocks along the fault slip, just as they do in large quakes, but they slip only a small distance—perhaps only a fraction of a millimeter. In terms of their violence, these tremblings are literally nothing compared with the immense New Madrid quakes, and yet it would be artificial not to refer to them as earthquakes. Like all earthquakes, they are sudden and abrupt slipping events that release energy within the Earth's crust.[21]

The truly fantastic difference between large and small earthquakes raises an obvious question: What is it that makes one quake big and another small? A quake that releases the energy of a thousand nuclear bombs clearly cannot be the same sort of event as one that is a hundred million times weaker. So what are the special conditions in the Earth's crust that set the stage for a really big earthquake? As we shall see, half a century of research has led scientists pondering this question to a disconcerting and rather paradoxical conclusion: There may be no such "special" conditions at all.

With reluctance, geophysicists are now beginning to accept that every earthquake, large or small, arrives at the far end of a long and immensely complex historical development within the Earth's crust. As a result, the dynamics of earthquakes can be understood only with the perspectives of historical physics, and especially through the concept of the critical state. The message: Massive quakes may arise out of the very same conditions as small, and quakes of all kinds may be totally unpredictable. As with avalanches in the sandpile game, the largest and most devastating earthquakes may take place when and where they do for no special reason at all.

· 3 ·

An Absurd Reasoning

To trace something unknown back to something known
is alleviating, soothing, gratifying and gives moreover a
feeling of power. Danger, disquiet, anxiety attend
the unknown—the first instinct is to eliminate these
distressing states. First principle: any explanation is better
than none. . . . The cause-creating drive is thus
conditioned and excited by the feeling of fear. . . .

—FRIEDRICH NIETZSCHE[1]

One of the chief services which mathematics has rendered
the human race in the past century is to put "common
sense" where it belongs: on the topmost shelf next to the
dusty canister labeled "discarded nonsense."

—ERIC TEMPLE BELL[2]

♦

IN THE POPULAR IMAGINATION, GREAT DISCOVERIES RESULT WHEN
solitary genius toils in obscurity and against all odds creates from the
rarefied atmosphere of pure thought some radical new idea. As one
myth goes, Albert Einstein lost patience with his teachers, sat down
in a patent office in Bern, Switzerland, and single-handedly turned
physics on its head. More often than not, however, great ideas have
prosaic origins. They are "born in a restaurant's revolving doorway,"
to borrow a phrase from Albert Camus,[3] or, in the setting of modern
science, when scientists from different fields meet and exchange
ideas.

So it was one morning in the summer of 1988, at a scientific conference held at a small college in New Hampshire. On this particular morning, geophysicist Yakov Kagan was giving a more or less
routine lecture on earthquakes, and, as most scientists attending
were not geophysicists, he was offering a general overview. Kagan
related the sad tale of the singular failure that he and his colleagues
continued to meet in trying to forecast earthquakes. And he also
introduced his audience to one of the few hard-and-fast laws ever
discovered about earthquakes, a rule describing how often earthquakes of various sizes take place. This rule is known as the
Gutenberg-Richter law. This law reveals that larger quakes take
place less frequently than smaller, and what's more, that the precise
numbers follow a relationship known to mathematicians as a *power
law*—a special mathematical pattern which has a simplicity that
stands in shocking contrast to the overall complexity of the earthquake process.

As chance would have it, Per Bak was sitting in the audience lis-
tening to Kagan's talk, and as Kagan spoke, Bak became increasingly
intrigued, because he and his colleagues had also found a power law

for the avalanches in their sandpile game. Could there be some peculiar similarity between earthquakes and the avalanches in a pile of sand?

Shifting Sands

In order to appreciate what excited Bak so much about the Gutenberg-Richter law, we should take a closer look at what it means. In the 1950s, seismologists Beno Gutenberg and Charles Richter, working then at the California Institute of Technology, hoped that a census of earthquakes over the whole earth might reveal some significant pattern that would provide a clue to the causes of quakes. Would they find, for example, some most common type of earthquake? Maybe earthquakes tend to have magnitudes around 3 or 7? Perhaps they only rarely have magnitudes such as 2, 5, or 8? Sifting through hundreds of books and papers, the two researchers assembled details about thousands of earthquakes. They counted how many had magnitudes between 2 and 2.5, between 2.5 and 3, and so on, and continuing this way they constructed a graph showing the relative frequencies of earthquakes of different sizes.

If there were such a thing as a typical earthquake, we should expect the graph to show one big hump—something like the famous bell curve of mathematics (SEE FIGURE 1). In this case, most quakes would fall at about some normal, average magnitude. Of course, one might instead find that earthquakes come not in one but in several different kinds, in which case we should find several humps in the graph. But Gutenberg and Richter found no humps whatsoever. They were looking at a worldwide catalog of earthquakes, but a more recent study for earthquakes in Alaska shows the pattern just as clearly (SEE FIGURE 2). Having no humps, the graph simply reveals that the bigger the quake, the rarer it is.

On the graph, I've plotted the number of earthquakes versus the magnitude. Recall that when the magnitude goes up by one, the

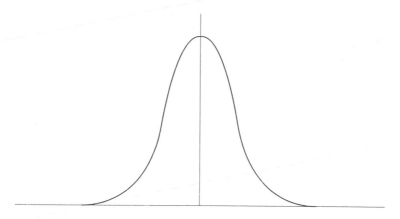

FIGURE 1. The bell curve is one of the most famous curves in mathematics. Weigh a thousand onions or apples, give a test to five hundred students, or measure the speeds of a few thousand cars as they rush by on the highway: in each case the numbers will fall on a bell-shaped curve, with the vast majority falling close to some average. If the statistics of anything follow the bell curve, then the numbers cluster together within a narrow range and finding any number far beyond that range is extraordinarily unlikely.

energy released in the quake goes up by ten. In terms of energy, it turns out that the Gutenberg-Richter law boils down to one very simple rule: If earthquakes of type A release twice the energy of those of type B, then type A quakes happen four times less frequently. Double the energy, that is, and an earthquake becomes four times as rare. This simple pattern—a power law—holds for quakes over a tremendous range of energies.[4]

Remarkably, Bak and his colleagues found a similar relationship for avalanches in the sandpile game. Counting up how frequently avalanches of each size happened, they found that avalanches toppling anything from a few up to a few million grains follow a regular pattern: Double the number of grains involved, and the avalanche becomes just a bit more than twice as unlikely (more precisely, about 2.14 times as unlikely). In the case of the sandpile, the researchers

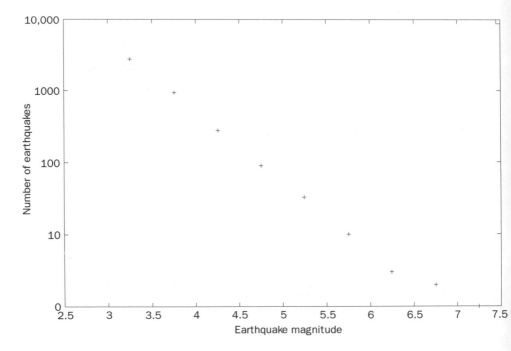

FIGURE 2. The energy released in any particular earthquake can vary over a terrific range. Even so, the statistics for all quakes reveal a strikingly simple pattern. Between 1987 and 1996, for example, a large number of earthquakes hit southern California in the United States. The points on the graph indicate how many there were within each interval of the earthquake magnitude: 2.0–2.5, 2.5–3.0, and so on. In terms of energy, the data indicate that earthquakes simply become four times less likely each time you double the energy they release. Data from the Southern California Earthquake Data Center, www.scecdc.scec.org/.

also had a good understanding of what lay behind this pattern. After the pile evolves to its critical state, many grains rest just on the verge of tumbling, and these grains link up into "fingers of instability" of all possible lengths. While many are short, others slice through the pile from one end to the other. So the chain reaction triggered by a single grain might lead to an avalanche of any size whatsover, depending on whether that grain fell on a short, intermediate, or long finger of instability. The power law simply reflects this situation, and points to the riddling instability that underlies the sandpile's workings.

In this simplified setting of the sandpile, the power law also points to something else: the surprising conclusion that even the greatest of events have no special or exceptional causes. After all, every avalanche large or small starts out the same way, when a single grain falls and makes the pile just slightly too steep at one point. What makes one avalanche much larger than another has nothing to do with its original cause, and nothing to do with some special situation in the pile just before it starts. Rather, it has to do with the perpetually unstable organization of the critical state, which makes it always possible for the next grain to trigger an avalanche of any size.

These details were running through Bak's head as he listened to Kagan's talk about the Gutenberg-Richter power law, and he began to wonder: Could things work the same way in the Earth's crust? If something similar were true of earthquakes, then there would be no essential difference—in terms of causes—between small earthquakes and the really big ones. They would all begin the same way and for the very same reason. This would immediately explain geophysicists' enduring troubles in making accurate predictions, or in identifying reliable precursors—for the idea that there are "special conditions" preceding the largest earthquakes would be a mirage. But can a game involving grains of sand really have any connection to the upheavals that can carve out new lakes and devastate entire cities? If the sandpile game is simple, the earthquake process is anything but. The idea of a connection was as preposterous as it was tantalizing.

On the other hand, Bak knew something else: that the stately regularity of the power law is not to be ignored. He knew that just about the only way to generate a power law pattern is by some process that is steeped in history—that is, in which the future emerges out of a string of accidents, each leaving its indelible trace on the course of events. A few years earlier, physicists Tom Witten and Leonard Sanders of the University of Chicago had given a dramatic illustration of this point while trying to understand something far simpler than earthquakes—namely, the nature of the freezing process. Suppose you melt some copper, and then cool it down so that it solidifies again. In 1984, Witten and Sanders invented a simple game to mimic the process, a game that shows quite clearly how a little history can give rise to power laws.

Let's take a closer look at the freezing game, for it reveals why it was by more than loose analogy with the sandpile game that Bak could conclude that large and small quakes must have similar origins. As we shall see, this game also reveals the key insight—concerning the importance of history—that allowed Bak to devise a game that would capture much of the truth about earthquakes.

Accidents of History

Toss a copper button into a pan of hot molten copper, and after a time there will be no button; it will have melted and its atoms will have dispersed into the liquid sea. After things settle down, the result is *equilibrium:* an unchanging steady state, in which the copper atoms swim about in unending, monotonous chaos. The liquid stays liquid, nothing interesting or different ever happens, and there is little to distinguish one moment of time from the next. In equilibrium, the notion of history has very little meaning.

But suppose you disturb the peace by taking the pan of hot copper and placing it into a bath of ice water. In an instant, some of the molten copper will cool and a chunk of solid copper will begin forming again in the pool. The copper is now *far from equilibrium*, since the liquid "wants" to freeze into the solid form, but hasn't managed it yet. Because of this imbalance, now there is such a thing as history, and the amount of solid copper will gradually increase with time. There is also something else—*complexity*. For as the atoms in the liquid fall over one another in trying to join the growing chunk of solid, they link up to form a weird and complex structure something like a snowflake, with all kinds of side-branches and whiskers. In a nutshell, this is what happens when something freezes very quickly: rushing from the liquid into the solid, the atoms pile up in an atomic-scale traffic jam, and a mess of complexity is the result.

In devising their game, Witten and Sanders hoped to come up with a way of simplifying this process so as to bring out the essential physics at work. To understand the game, forget about real copper for the moment, and just imagine a single particle that starts out at the center of an empty space. From afar, another particle wanders in, following a random, haphazard path. If the second particle misses the first, then it just keeps going. But if it hits, it sticks. Next, a third particle comes in, again following a random path. Its fate is determined in the same way. If it hits any particle in the center cluster (which might now be one particle or two), then it sticks; if it misses, it just keeps going. The game is to keep sending in particles at random, letting them either stick or miss, and see what happens.

Even though these rules are so simple, the cluster that forms has a fantastic shape (SEE FIGURE 3)—much like the real clusters of copper that form during freezing. And the reason for the complexity is not hard to see. In the game, when a particle hits the cluster, it stays there. This is a definitive historical happening, irreversible in its consequences, and it leaves its traces on everything that happens later. The sticking of each particle changes the shape of the cluster and makes it more likely that other particles will stick near the same place. When they do, this makes it even more likely that still more particles will stick. The growth is highly unstable, and every accident leaves its indelible trace in the growing structure forevermore.

So in this far-from-equilibrium setting there is such a thing as history, and it is a very important thing indeed. It is what makes the clusters complex rather than simple. This is the first lesson of Witten and Sanders's game. But there is another lesson too. If you grow

FIGURE 3. In the aggregation game, sticky particles wander in randomly from afar and attach themselves if they happen to hit the growing cluster. After a long time, the result is something like this picture. Run the game more than a million times and you'll never find quite the same structure—in its precise details, the result is utterly unpredictable. Even so, it always has a similar appearance and satisfies precise power laws. Image courtesy of Paul Meakin, University of Oslo.

clusters a hundred times, every one will be unique. This is another expression of the complexity of the clusters, each of which reflects a long string of historical accidents. And yet these clusters also possess a hidden order.

Suppose you decide to inspect some cluster in detail, and by a stroke of good fortune you are a being who can change your size at will. You might begin at the size of a cherry, look around for a while, and get a feel for your surroundings. You notice some branches that are roughly cherry-sized and others that are somewhat larger or smaller. Perhaps then you decide to shrink yourself down to a size ten times smaller. You would be surprised to find that things at this smaller scale look much the same. As it turns out, the landscape of the cluster at all scales has precisely the same feel to it, and if you lost track of how many times you had shrunk yourself, you wouldn't be able to tell merely by looking around.

Somewhat bewildered, you might decide to try and put some numbers on the features of this strange landscape, in the hope of learning a bit more about its general character. To do so, you could proceed à la Gutenberg and Richter—counting up how many branches you find of each of the various sizes from large to small, and then making a graph. You would find a power law: Every time the size of the branches decreased by two, the number of such branches would increase by about three. You could start from scratch and run the game again, and you would again find the same result—the clusters *always* have this property. Even within the storm of confusing accidents out of which the cluster grows, and even though the future is at every moment turned one way or another by random chance, a remarkable order still emerges.

So the second lesson of Witten and Sanders's game is that while history can lead to complexity, it can also lead to a special, hidden simplicity—for the clusters always have a property known as *scale-invariance* or *self-similarity*. As you discovered when making yourself different sizes, the cluster looks much the same at all scales. Its pieces are rough copies of the whole—hence the term "self-similar." This spectacular feature shows up mathematically in the power law for the distribution of branches by size. This, then, is what the power law means—it is a mathematical signpost telling you that the cluster looks the same at all scales.

Another way to say the same thing is that in the freezing game, there is no one scale that is "preferred" over another. Chickens never lay eggs as big as a basketball or as small as a dust mite; the design of a chicken gives it a built-in bias toward producing eggs falling on a bell curve around a familiar, typical, normal size. But in the freezing game, there is no bias; the rules are naturally suited to producing branches over a tremendous range of scales. And this brings us back to earthquakes.

One Cause

Bak knew all about this freezing game, and he knew that if the distribution of something follows a power law, then words such as "normal," "typical," "abnormal," and "exceptional" simply do not apply. Whether they describe avalanches in a sandpile, branches in a cluster, or anything else, power laws always have this implication. Consequently, the Gutenberg-Richter power law shows that the process behind earthquakes is scale-invariant, and the unavoidable implication is that the great quakes are no more special or unusual than the tiny shudders constantly rippling beneath our feet.

Of course, the power law doesn't explain *why* earthquakes work this way; it just says that they do. But the freezing game reveals that, in one sense or another, history seems to be important in the formation of power laws, and that one way to study things in which simplicity and complexity mingle is by building rudimentary historical games—or, in the language of physics, games for processes that are far from equilibrium. All of this was swirling in Bak's head as he left the conference at which he had heard Kagan speak. And back at Brookhaven, he and his colleague Chao Tang set to work.

The two researchers began talking with theoretical geophysicists and rummaging through the research literature, looking for hints on how to build a game that would use the lessons of the Gutenberg-Richter law and would capture the essence of the earthquake process. Might they devise some game, like the freezing game, that would reveal the underlying processes behind the triggering of earthquakes? At first, unfortunately, they could find little about the mechanics of earthquakes that seemed in any way simple. The San Andreas Fault, for example, is indeed a line that cuts through California from north to south, yet near the main fault the Earth's crust is riddled on both sides by thousands of smaller faults, and from each of these spring still smaller "subfaults," and so on. California is as run through with cracks as the concrete of any worn-out and dilapidated roadway, and the word "fault" really ought to be replaced by the phrase "fault system." Every earthquake zone on Earth shares this same fractured character. In China, Colombia, California, or Japan, the faults and subfaults and smaller cracks form their own peculiar networks, some cutting through mountainous terrain, others carving up plains or rolling hills, and others slicing through crust that lies on the ocean floor.

Facing this mass of detail, Bak and Tang were making little progress, until one day they got lucky. In an old paper from 1967, Bak and Tang read how geophysicists R. Burridge and Leon Knopoff of the University of California at Los Angeles had tried to strip away most of the confusing complexity of the physics behind earthquakes in the hope of gaining at least some insight into their causes.[5] The game they had arrived at was wildly unrealistic, and yet for Bak and Tang, it was a place to start.

Burridge and Knopoff had aimed to ignore almost all the detailed complexity of real fault systems, and so chose to focus on just a single fault instead. Recall that along the principal fault of the San Andreas, the plate to the west drifts northward, while the plate to the east moves southward. If the rocks don't slip at the boundary, then they bend, and develop internal stress. The more stress, the more likely slipping becomes. Earthquakes come from the buildup and release of stress. Burridge and Knopoff tried to picture schematically how it might happen.

Imagine a great wooden floor and, above it, an equally great ceiling that moves inexorably like a conveyor belt to the right (SEE FIG-

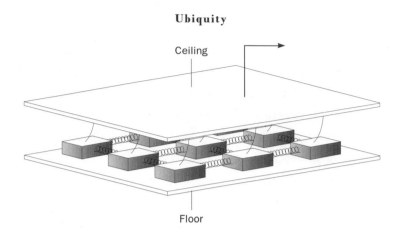

Ceiling

Floor

FIGURE 4. The Burridge-Knopoff model for earthquakes. Adapted from Z. Olami, H. J. Feder, and K. Christensen, "Self-Organized Criticality in a Continuous, Non-conservative Cellular Automaton Modeling Earthquakes," *Physical Review Letters* 68 (1992): 1244–1247, reprinted by permission.

URE 4). The ceiling has some flexible rods attached to it, which run down to a bunch of wooden blocks on the floor. The game works like this: As the ceiling moves, the rods bend, and so try to drag the blocks along. Friction with the floor tends to hold blocks in place. The more the ceiling moves, the more the rods bend. For each block there ultimately comes a time when friction is defeated, and the block suddenly slides forward.

One further detail makes the game interesting: some springs connected between the blocks. Without these, the blocks would move independently of one another, and would simply stick, slip, stick, slip, and so on. But with the springs, when one block slides forward, it pushes on the one in front of it, and pulls on the one behind. Also, if for any reason a block moves off to one side, it pushes and pulls on the blocks to either side. The total force trying to make any block slip now comes not only from the flexible rod overhead but also from the springs before and aft, and to its left and right. The springs let the motion of one block influence that of another.

What does this have to do with earthquakes? The idea is simple. The game is supposed to represent a geological fault with all the real life sucked out of it, and only the essential physics remaining. The floor and ceiling represent two continental plates, and the surface between the blocks and the floor represents the two immense surfaces of those plates. As long as friction keeps the rocks from slipping, they can only bend and store up energy. The flexible rods capture this. And the springs between the blocks? You or I cannot compress a slab of granite in our hands like a spring, and yet when two continents grind together, they have little trouble doing it. Suppose there is an earthquake. If the rocks at different points along a fault slip by unequal amounts, then some regions of rock will be compressed, and will want to spring back. The springs represent—schematically, again—the fact that rocks are elastic.

The game, then, is to just let the ceiling drift, and see what happens. Burridge and Knopoff had built a simple physical version using a one-dimensional string of blocks, rather than a two-dimensional network. Unfortunately, they could work the device only in a fairly rudimentary way, with a small number of blocks, and looking at only a small number of slipping events. That was in 1967. In 1988, Bak and Tang could use an ordinary desktop computer to study the same thing. And to their amazement, the computer let them discover something that Burridge and Knopoff had only dreamed of.

The Hinge of Fate

To simulate the game on a computer, it is necessary to know some further details. Each block is supposed to slide when the forces on it pass some threshold. But once it begins sliding, what happens next? How far does the block go? The movement of a block should properly be described by Newton's equations of motion, including the effects of friction between the block and the floor. But to keep things really simple, Bak and Tang decided to replace Newton's laws with an ad hoc rule: If the force on a block exceeds the threshold for slipping, then that block will step forward a fixed distance. At the same time, the force on that block will decrease by one unit, and the force on each of its four neighbors will increase by one-fourth of a unit. In other words, when a block slips, some of the force that was acting on it gets shifted and divided up equally among its neighbors.

Bak and Tang had no real justification for their rule, but one of its great advantages was that it made the model so simple that the researchers could run simulations with millions of blocks very quickly. Ironically, however, they never had to. For even before they got started, they had a sudden attack of déjà vu. Without intending it, their ad hoc rule had bestowed upon the earthquake game a mathematical logic that was identical to that of the sandpile game. Here the language referred to blocks slipping over a surface; there to grains of sand toppling down a pile. Yet behind the language was just one mathematical skeleton. So Bak and Tang didn't even have to play this new game to see how it would work. They had found a far-from-equilibrium game for earthquakes, all right, and it was simply the old sandpile game in disguise.

Having got this far, Bak and Tang could finally broach the ultimate question: If avalanches in the sandpile game follow a power law, what does this say about earthquakes, at least earthquakes of the block-and-spring variety? The Gutenberg-Richter law refers to the number of quakes that release a certain amount of energy. It was fairly simple to see how this would translate into the earthquake game. Bak and Tang reasoned that each time a block slides forward one step it should release roughly the same amount of energy. So the total "energy" of a quake in the game is simply the total number of block-sliding events triggered by one initial event. If we compare the two games, we find that this is also the total number of grains that topple, or the size of an avalanche triggered by the dropping of one grain. So, by analogy, quakes in the earthquake game follow the same power law as avalanches in the sandpile game. Something very much like the Gutenberg-Richter law for real earthquakes comes tumbling out.[6]

To Bak and Tang's excitement and astonishment, their simple game seemed to explain and predict the most basic law of all earth science. One might raise a minor objection: double the size of an avalanche in the sandpile game, and it becomes 2.14 times as rare. The Gutenberg-Richter law, in contrast, says that when the size of an earthquake doubles, it becomes 4 times as rare. So the numbers in the power laws aren't quite the same. And yet this small discrepancy seems rather insignificant when compared to Bak and Tang's achievement: they had at least explained where the special form of the power law comes from.[7]

The game also had the right feel to it. A recording of how many blocks were slipping at each moment during one play of the game revealed a bizarre but strangely familiar pattern. The huge earthquakes stood out like redwoods in the desert, and to the unsuspecting eye appeared to demand some special explanation—just as large earthquakes do to geophysicists. Yet the initial cause of each, large or small, was just the sudden slipping of one block somewhere in the system. Since the blocks and springs organize themselves into the critical state, the system is always balanced uneasily, and anything can happen. The slipping of one block might trigger a system-wide avalanche of slippings—that is, a catastrophic earthquake. The difference between the enormous and the small has only to do with the precise location of the very first slipping event. This is why earthquakes are unpredictable, and why terrible upheavals can strike without any warning at all.

These results seemed almost too good to be true—and they were. Other researchers immediately objected to Bak and Tang's pared-down version of the game, and their ad hoc rule came in for a particular savaging. The sandpile game has the property of being "conservative:" as grains topple down the pile, their total number is conserved. Grains never disappear. In their earthquake game, Bak and Tang had unwittingly made the forces on the blocks work in the same way. When the force on one block became large enough for the block to slip, their ad hoc rule then distributed this force in equal shares to that block's neighbors. So the total force trying to make blocks slide stayed the same, as does the total number of grains in the sandpile game. That's why the two games ended up being the same.

Trouble is, the real world isn't like this. If geophysicists don't know much about the laws of friction between sliding rock surfaces, they do know for sure that there *is* friction—and this friction ought to eat up some of that force. So the Bak-Tang rule is most certainly not legitimate. Trying to respond to this objection, Bak and Tang altered the rule to make things more realistic. But when they did, the game changed into something else. It wasn't equivalent any longer to the sandpile game, and its power law evaporated. Earthquakes were not explained.

Mistaken Identity

Just about a year later, in 1990, while still searching for a better solution, Bak and another colleague, Kan Chen, wrote a long article about earthquakes and the critical state and sent drafts of the manuscript to other physicists, including Jens Feder in Oslo, Norway. Together with his son Hans Jacob—then still in high school—Feder began running some computer simulations to try and replicate the results of Bak and his colleagues. The Feders wrote a program following the rules set out in the paper, and began playing the game. They recorded how often earthquakes of different sizes occurred. And eventually, having acquired enough data, they plotted it. As expected, they found a power law. Disturbingly, however, it wasn't the same power law that Bak and his colleagues had found. The precise numbers were different.

The Feders checked their computer code, and rechecked Bak and Chen's description, but everything seemed correct. They ran more simulations, and still the altered power law persisted. In frustration, Jens Feder finally got Bak on the phone and after long discussion flushed out the cause of the trouble: that early draft of the paper contained a tiny but crucial misprint. One of the rules of the game had been written down incorrectly. So the Feders had been playing the game with the wrong rules. Indeed, they were playing a different game. Miraculously, however, this new game wasn't senseless or boring. If great ideas can be born in "a revolving doorway," then important discoveries can arise from typographical errors, for this new game still gave rise to behavior much like that described by the Gutenberg-Richter law, and yet didn't suffer from the problem of being conservative—that is, it didn't automatically flout the known principles of friction.

The next summer, Hans Jacob Feder teamed up with Zeev Olami and Kim Christensen, associates of Bak's at Brookhaven, to try to understand how this could be. Returning to the original game of Burridge and Knopoff, the trio began retracing Bak and Tang's footsteps, paring down the game, simplifying, yet trying to remain faithful to the underlying physics. Soon they came to the sticky point of how to specify the movement of a block after it has begun sliding. But where Bak and Tang had introduced their troublesome ad hoc rule, Olami, Feder, and Christensen found another way around it. And this other way had a solid grounding in physics.

Rock sliding against rock makes heat. That is, some of the energy stored in the stressed rock goes not into moving the rock, but into heating it. To mimic this, the game needed to include a similar mechanism for dissipating energy. Olami, Feder, and Christensen easily invented a new rule to do just that. If a block slides forward, they proposed, it should lose some energy owing to friction with the floor. So while the force on the slipping block should decrease by a unit, the forces on its neighbors should increase by less than this amount. That was all it took. With the new rule replacing the old in Bak and Chen's game, the Olami-Feder-Christensen game—originally discovered because of a typographical error—came tumbling out.[8]

As a bonus, this game fit the facts about earthquakes even better than the original Bak and Tang game. To begin with, in the Olami-Feder-Christensen version, each doubling of quake size leads to the quake's becoming four times as rare—precisely the same value seen in the Gutenberg-Richter law for real earthquakes. Moreover, in 1995 physicist Keisuke Ito of Kōbe University in Japan ran extensive simulations using a slight variation of the game, and looked at the precise times when earthquakes struck.[9] Earthquakes in the real world tend to be accompanied by foreshocks and aftershocks, which is another way of saying that large earthquakes tend to cluster together in time, and that the longer you wait without seeing one, the longer you will probably have to wait. Ito found that this clustering also falls quite naturally out of the game.[10]

How Can It Be Like That?

Through the ideas of Bak and Tang; Olami, Feder, and Christensen; and many other researchers I have not mentioned, earthquakes in recent years are finally beginning to be explained. Or, more accurately, the character of the process that lies behind them is no longer a deep mystery. The American physicist Richard Feynman counseled students of quantum theory to avoid falling into an intellectual abyss by asking, "How can it be like that?" Fortunately, there seems to be no danger of this in the case of earthquakes. But understanding "how it can be like that" is not the same as being able to make predictions. Indeed, in this case understanding leads instead to the conclusion that prediction of individual earthquakes is probably impossible, even though the basic picture of the earthquake process is quite simple.

The Earth's crust is under constant stress owing to the motion of plates, driven to move about by heat in the Earth's interior. This stress builds up until the rock along one tiny segment of a fault reaches its threshold for slipping, and slips. This initial segment might be only a millimeter long. It might even be microscopic. But what happens next needn't be, for the magnitude of the ultimate effect bears no relationship to that of the initial cause.

If the Earth's crust works like the earthquake game, or its close cousin the sandpile game, then over time the stress and strain in the various bits of rock acquire the special organization of the critical state. The crust becomes riddled by fingers of instability of all possible lengths. So after the first bit of rock slips, somewhere, quite literally anything might happen. The earthquake may stop quickly. Or the initial movement may place enough stress on neighboring pieces of rock to cause further slipping. The ultimate size of an earthquake depends on one very tiny detail that perhaps lies forever beyond our scrutiny: the length of the particular finger of instability on which the first tiny slipping event takes place.

Catastrophic earthquakes, then, strike in a very real sense for no reason at all. There is an explanation for why there are such earthquakes in the first place: it is the very fact that the Earth's crust is tuned to be in a critical state, and lives on the edge of upheaval. But there is no explanation—other than a simple narrative telling, after the fact, of which rocks slipped and in what sequence—for why a quake such as the New Madrid earthquake of 1811 was so big. The first rock that slipped just happened to lie on a very long finger of instability. These fingers run through all fault systems. A huge quake could happen at any time in any of them. In the picturesque phrase of the earthquake expert Christopher Scholz of Columbia University, it seems that an earthquake when it begins "does not know how big it is going to be." And if the earthquake itself doesn't know, we aren't likely to know either. Clues as to what causes any one earthquake seem not to lie just in the region where the quake took place; rather, understanding the earthquake process means coming to terms with the complex organization of the ever changing pattern of stresses and strains in the Earth's crust as a whole.

You might suppose, of course, that if you could take an accurate picture of the pattern of stress and strain, and learn in incredible detail the properties of all the rocks, how much stress they can withstand before slipping, and so on, then perhaps you could make a map of the fingers of instability. But even then, prediction of the large quakes would be next to impossible. There must be hundreds of millions of places in the crust where tiny microscopic slipping events are just about to happen because the rocks there have been pushed to their limit. You would need to monitor all those points to be sure that none fell on one of the longer fingers of instability.

World Out of Balance

This book is not only about earthquakes. It is about ubiquitous patterns of change and organization that run through our world at all levels. I have begun with earthquakes and discussed them at some length only to illustrate a way of thinking and to introduce the remarkable dynamics of upheavals associated with the critical state, dynamics that we shall soon see at work in other settings. When it comes to disastrous episodes of financial collapse, revolutions, or catastrophic wars, we all quite understandably long to identify the causes that make these things happen, so that we might avoid them in the future. But we shall soon find power laws in these settings as well, very possibly because the critical state underlies the dynamics of all of these different kinds of upheaval. It appears that, at many levels, our world is at all times tuned to be on the edge of sudden, radical change, and that these and other upheavals may all be strictly unavoidable and unforeseeable, even just moments before they strike. Consequently, our human longing for explanation may be terribly misplaced, and doomed always to go unsatisfied.

· 4 ·

Critical Thinking

The aims of scientific thought are to see the general in the
particular and the eternal in the transitory.[1]

—ALFRED NORTH WHITEHEAD

_____

The purpose of models is not to fit the data,
but to sharpen the questions.

—SAMUEL KARLIN[2]

◆

IN THE EARLY AFTERNOON OF DECEMBER 2, 1942, A TEAM OF PHYSICISTS
filed down the stairs leading to the squash courts underneath the
football stadium of the University of Chicago. It was the occasion of
a historic experimental test. In a makeshift laboratory set up in one
of the courts, the team had built what was meant to be the world's
first nuclear reactor. They had drilled holes into an enormous block
of graphite and inserted long rods of enriched uranium. Enrico
Fermi, the project's mastermind, had discovered six years earlier that
uranium nuclei, when hit by a neutron, could be made to split apart
and release further neutrons. These, slamming into other uranium
nuclei, could in principle trigger still further splittings and an
avalanche of further neutrons—a self-sustaining nuclear reaction.

That was the theory, at least. The theory also held that unless
one was careful, the reactor would kick into gear all by itself. After
all, even without prodding, a uranium nucleus disintegrates every so
often, and emits a neutron. Under the right conditions, even a single
such neutron could trigger a runaway chain reaction. To keep his pile
from going off before he was ready, Fermi had inserted "control
rods" made of cadmium in among the uranium fuel rods. These kept
the brakes on the pile by gathering in neutrons and ensuring that the
avalanche triggered by a single neutron would soon die out. But on
this day, Fermi was ready to take the brakes off and see what would
happen.

A little after 3 P.M., everyone took a deep breath as Fermi
grabbed hold of a rope and began sliding the cadmium rods ever so
slowly out of the block. The physicist Eugene Wigner stood nearby
with a celebratory bottle of Chianti, nervous but hopeful. As the rods
inched out, a Geiger counter began registering the occasional click;

a bit further, and it began rattling like a machine gun. Fermi had calculated with his slide rule how far he could go before the pile would run away into a catastrophic chain reaction, and at 3:36 P.M., as he approached that point, the Geiger counter went crazy. Fermi stopped pulling the rods out. He had tuned the pile to within a whisper of the critical point, where a single neutron could trigger an avalanche of any size.

The message in this story is that nothing reaches the critical state all by itself. To place something on the very edge of instability, and to keep it there, usually requires careful tuning and continual adjustment. And this is why, in 1987, Bak, Tang, and Weisenfeld had been puzzled, mystified, and astounded that their simple sandpile game seemed to develop into the critical condition quite naturally. Starting with a flat surface, their computer dropped grains slowly and at random. The pile grew, became steeper, and then the avalanches began. At first, they involved only a few grains. As the pile grew, so did the typical size of the avalanches. Ultimately, the pile entered the critical state and was susceptible to avalanches of all sizes, as had Fermi's pile when he tuned it properly. But Bak, Tang, and Weisenfeld hadn't tweaked any knobs to bring it there. The critical organization welled up on its own.

Recognizing a miracle when they saw one, they enshrined it with the name "self-organized criticality." And the miracle trailed a fascinating possibility in its wake. If in the sandpile game the spectacular organization of the critical state seems to arise completely for free, with no tuning needed, might other things in nature work the same way? We already know that self-organized criticality appears to account for the unruly and unpredictable workings of the Earth's crust, which seems to live in a critical state. The slow, inexorable drifting of continental plates is something like the dropping of grains, and has brought the crust into a state in which "avalanches"— in this case involving the rocks slipping past one another along fault lines—come in all sizes. Are there other things in the world, things that seem just as complicated, but which also share the essential logic of the sandpile game?

For more than a decade this question has been surrounded by a "hectic air of controversy."[3] Physicists still haven't nailed down all the answers, but what they have found is as subtle as it is fascinating.

Burning Properly

Understanding what made the great 1988 forest fire in Yellowstone National Park so terrible is far from easy. Why and how and where fire spreads depends on the kind of trees in its path, on how far apart those trees are, and on the more detailed patterns in which forest and grassland mingle. Winds drive a fire to spread, while rain slows it down. The detailed history of the forest matters too; growth in some regions is much older than in others, and this affects how easily it burns. Natural barriers such as rivers can retard the advance of a fire; then again, a hot fire can blow embers clear over a river and set new fires more than a mile ahead.

Given all these influences, it is perhaps no surprise that scientists have about as much success in predicting large forest fires as they do in predicting earthquakes. If the U.S. Forest Service was caught completely off guard in Yellowstone, that may be because there were simply too many details to take into account. Then again, there may be a deeper reason. In 1998, the geologists Bruce Malamud, Gleb Morein, and Donald Turcotte of Cornell University gathered extensive data on forest fires in the United States and Australia over the last century. The size of a forest fire is sensibly given by how many trees it burns, or, equivalently, the area that the fire consumes. How large is a typical forest fire?

One would expect that the history of fires might reveal some rough stalemate reached between the destructive forces of nature and the preserving efforts of mankind. To find out for sure, Malamud and his collaborators made simple graphs showing how often fires consumed 1 square kilometer, 10 square kilometers, and so on. Surprisingly, they did not find any indication that there might be a typical size for a fire. For example, their data for 4,284 fires on U.S. Fish and Wildlife Service lands between 1986 and 1995 reveals a remarkably strong power law. Once again we find the same geometric pattern: double the area covered by a fire, and it becomes about 2.48 times as rare, and the pattern holds for fires varying in size by a factor of a million. In other words, despite the immensely complex picture of how fires spread, a startlingly simple pattern emerges when you look at how often you find fires of different sizes—a kind of Gutenberg-Richter law for ecological conflagration.

Recall that a power law, with its scale-invariant form, implies that large events are just magnified copies of smaller ones, and that they arise from the same kinds of causes. Really big earthquakes aren't triggered by special events, but are simply the natural if infrequent consequence of the overall critical organization of the Earth's crust, and its susceptibility to long-range chain reactions. The Cornell researchers found that the same thing seems to be true for forest fires, not only in the United States, but also in Australia, and presumably everywhere on Earth. When a fire starts, it doesn't yet know how big it will become. Fires spread as they do because any forest has the organization of the critical state, and how far any particular fire goes is largely a matter of chance.

That, at least, is what the raw data suggest. Of course, a power law by itself is only data. It may hint at the lack of any distinction in the causes of large and small earthquakes, but a skeptic might still wonder. To gain some deeper insight into how such a power law might arise, Malamud and his colleagues went one step further. We have already seen how Witten and Sanders had managed to capture the essence of freezing under far-from-equilibrium conditions in their freezing game; Bak, Tang, and other researchers managed the same thing using simple games for earthquakes. Malamud and his colleagues aimed to follow in their footsteps, and to devise a game that would reveal the essence of the process, while not getting bogged down by too many details. So what is essential about forest fires?

Insofar as the spreading of fire is concerned, the Cornell researchers boiled things down to three principles. First, forests are made of trees, and, left to themselves, these trees will with time increase in number. Second, once in a while, some tree somewhere will catch fire. Third, this fire will spread to other trees nearby. To forestry workers, this skeletal picture represents a ludicrous over-simplification of a real forest. Nevertheless, Malamud and his colleagues put these principles into a mathematical game, and then used the computer to see how it might work.

Like the sandpile game, the forest fire game is played on a grid, and, at each time step, the computer plants a tree on a random square. As time runs on, the number of trees increases as they sprout up at random all over the forest. Every so often, however, after a certain number of trees have been planted, the computer drops a match on a random square. So, we have trees popping up at a uniform rate, one each step, and matches falling on squares at a smaller rate—say, one match for every two hundred or four hundred trees that sprout. When a match falls, it does nothing if it lands on an open square. If it hits a tree, that tree catches fire. The final rule in the game is that, once a tree catches fire, it will at the next time-step set fire to any trees that happen to occupy one of the four squares next to it. That is all. The game grows trees at random, sets single trees on fire once in a while, and lets the fire spread, if it can.

Malamud and colleagues ran a number of simulations, and in each they counted how many times they saw fires that burned off a given area of the grid. There were, as in real forests, many more small fires than large. But beyond the mere qualitative agreement, the model also gave rise to a near-perfect power law.[4] The network of trees on the grid seemed naturally to tune itself to a critical state in which the next match might spark a fire of any size whatsoever, even one that would destroy the entire forest.

Based on the remarkable correspondence with their simple game, Malamud and his colleagues conclude that the Earth's crust isn't the only thing to organize itself into a critical state; forests do it as well, at least if left more or less to themselves. This qualification is necessary. For the game also turned up one other curious detail, one that may even help the U.S. Forest Service to reduce the number of huge, catastrophic fires in the future.

Supercritical

The 1988 fire in Yellowstone burned 1.5 million acres. In a critical state, of course, there is no reason to look for specific causes of really big events. The mere existence of the critical organization means that terrific fires will occasionally break out, no matter what, since the forest is poised on the edge of disaster, just like Fermi's critical reactor. But it now appears that the forests in Yellowstone and other U.S. parks may be in an even worse situation. If Fermi had not stopped pulling out his control rods, the reactor would have fallen into a disastrous runaway reaction, with each neutron triggering an avalanche involving an ever growing number of others. No one would have celebrated that day with Wigner's Chianti. Unfortunately, U.S. forest management policy over the past century may have committed the ecological equivalent of pulling the control rods out all at once, with the result that the forests are now not merely on the edge of disaster, but anchored firmly in its almost certain path. The game shows why.

Recall that the computer dropped a match every so often. Malamud and his colleagues could alter how often. In some runs, they made the computer drop a match after each one hundred trees that were planted; in others, they made it drop one only after every two thousand trees went in. In the first case, matches fall fairly frequently, and there are many fires. In the second case, the matches fall ever more rarely, and so fires flare up less frequently. And what happens in this second case is instructive: since there are not many fires, the density of trees tends to become high, as nothing ever gets rid of them. When two thousand trees were planted for every match dropped, in fact, the game typically filled up the entire grid with trees before a sparking a single fire. When it finally did, the result was as catastrophic as it was inevitable: a single tree torched a fire that spread across the entire forest. In other words, when the fire-starting frequency was very low, the game showed a marked tendency to catastrophic, all-consuming disasters.

Malamud and his colleagues dubbed this the "Yellowstone effect," and it may explain why the U.S. Bureau of Land Management has acknowledged that, despite determined efforts to suppress naturally ignited fires, wildfires have in recent years become more numerous, severe, and difficult to control. From 1890 onward, the attitude of the U.S. Forest Service was one of "zero tolerance," even for forest fires sparked by natural causes. The service tried desperately to put out every fire whatsoever. This is the real-world equivalent of dropping matches far less frequently in the forest fire game, and it appears to have had similar consequences.

One of the unintended effects of this program was that the forests began aging.[5] Old trees were not replaced by younger trees, and the natural evolution of the forest's materials changed. Deadwood, grass and twigs, brush, bark, and leaves accumulated; as a result, the forests moved away from the natural critical state. The trouble is that fires are an indispensable component of the natural dynamics that keep forests in that state, so by suppressing them, the Forest Service has instead driven the forests into an even more unstable state, a supercritical state, with a high density of burnable material everywhere. Mother Nature has, as one writer has commented, "hidden the equivalent of a doomsday device" in the forest. "The protected woods have built up an enormous fuel load of downed and standing dead trees and limbs, flammable underbrush and grass . . . a single lightning strike or cigarette butt can explode into mass fire."[6]

The U.S. Federal Wildland Fire Policy now recognizes the difficult position into which U.S. forests have been put by past practices. It states:

> Catastrophic wildfire now threatens millions of wildland acres, particularly where vegetation patterns have been altered by past land-use practices and a century of fire suppression. Serious and potentially permanent ecological deterioration is possible where fuel loads exceed historical conditions.[7]

Consequently, forest managers are no longer trying to control the small and intermediate-size fires. Indeed, they now even set prescribed and managed burns in order to keep the fuel from building up. Fires of intermediate size remove some of the dangerous deadwood from the forest. In analogy with the forest fire game, they reduce the density of paths along which fire can spread, and so make it more difficult for a tiny disturbance to trigger large-scale disaster. The U.S. Federal Wildland Fire Policy hit the nail squarely on the head in concluding that "wildland fire, as a critical natural process, must be reintroduced into the ecosystem." It may take years to redress the balance, and even then large fires will, of course, still break with a fair frequency; this is unavoidable in the critical state. But at least terrific conflagrations would be less likely than they would be in the supercritical situation.

The forest fire model indicates that the essential truth about how fires spread has little to do with the details of the things involved. The forests appear to be an excellent example of self-organized criticality. What counts in the critical state are not complex details but extremely simple underlying features of geometry that control how influences can propagate.

Relatively Critical

Early in the twentieth century, thinkers of all sorts pointed to Albert Einstein's theory of relativity as proof that the truth about anything depends essentially on perspective. It is ironic that for Einstein the theory offered precisely the opposite message. Relativity theory is based on the notion of invariance—that is, on a deep appreciation of things that remain the same even when one makes a change of perspective. The idea of self-organized criticality shares this spirit, and herein lies its power. It is a one-size-fits-many explanation for the workings of things, irrespective of the myriad bewildering details of the molecules, trees, or what have you that make up those things.

In western North America, more than three hundred different species of grasshoppers graze in the flat grasslands. Nothing that lives above ground consumes more vegetation. In a typical year, these grasshoppers eat about 20 percent of all available leafy foodstuffs and, in so doing, dramatically affect the grassland ecology. Ordinarily this is good, as their foraging helps to cycle nutrients through the soil and to maintain the stability of plant communities. Sometimes, however, their numbers can get out of hand. In 1983 and 1984, in the Black Hills region of Wyoming, grasshopper numbers soared and the grasslands were virtually smothered. Authorities have been trying to anticipate and control similar outbreaks all over the American West for more than a century, since ranchers depend on these lands for grazing livestock. However, trying to understand what makes for a dramatic outbreak is anything but easy. Ecologists estimate that more than twenty thousand factors, including the detailed pattern of seasonal temperatures and rainfall, and the populations of numerous grasshopper predators and parasites, go into determining how grasshopper numbers will change from one year to the next.

In 1994, however, the ecologists (and brothers) Dale and Jeffrey Lockwood set out to study the outbreaks in greater mathematical detail than had been done before, and what they found may no longer surprise you. In various regions of Idaho, Montana, and Wyoming, the U.S. Department of Agriculture has for more than half a century recorded year-by-year data revealing the total land area on which grasshopper numbers have exceeded a certain threshold known as the carrying capacity. Roughly speaking, when the density of the grasshopper population exceeds this value (about eight per square meter), the grasshoppers' activity in just one year can leave long-lasting scars in the structure of the local plant community. The area covered at such a level offers a good measure of the size of a grasshopper infestation. Looking at the outbreak records for several regions, the researchers found a distribution in size conforming to a power law. If small outbreaks are common, larger ones are not. The important point, however, is that there appears to be no meaningful distinction between the two in terms of their causes. The power law suggests that the apparently insignificant causes that trigger a small outbreak on one occasion may initiate a devastating outbreak on another, and that no analysis of the local conditions at the initiation point of an outbreak will suffice to estimate the outbreak's ultimate extent.[8]

For pest management, this may provide a lesson similar to that learned in the forests. On one hand, it suggests that even if ecologists had a perfect handle on all those twenty thousand factors, they would be no better at making outbreak predictions. Large outbreaks may be inevitable simply because the dense web of interactions between organisms and physical influences lives in something like a critical state, poised on the edge of dramatic change. Trying to predict the next outbreak is therefore probably senseless. On the other hand, it may be that past efforts to suppress outbreaks were misguided, in that this very suppression may only serve to increase the likelihood of a great outbreak.

A similar pattern holds in other settings as well. In 1996, two researchers at the University of Oxford, Roy Anderson and Chris

Rhodes, studied measles epidemics between 1912 and 1969 in the isolated human population of the Faeroe Islands, situated in the North Atlantic between Iceland and Norway. They discovered that the distribution of measles epidemics according to size (number of people infected) follows a beautiful power law just like that for earthquakes or forest fires. Simulations with the forest fire model with the trees representing people, and fire representing infection, accounted accurately for the observations.[9] Again, the details just don't seem to matter. A model invented to account for the spread of fires in forests, and which does so extremely well, also captures the essence of how diseases spread though some human populations. Even if the "trees" are people, and the "fire" is a disease, disturbances still spread in the very same way.

The critical state may also explain the workings of some of the more bizarre objects in the universe. The stars known as pulsars are made entirely of neutrons. They are outrageously dense: a mere teaspoonful of the material would far outweigh the largest skyscraper. Pulsars are also astronomical lighthouses; each sends out a beam of light that swings through space as the star rotates. Astronomers detecting the light on Earth see a pulse every time the Earth passes through the beam. Every so often, however, the pulsing rate of a pulsar suddenly jumps up a notch, as if the star had suddenly started spinning faster. These jumps are known as pulsar glitches.

Some glitches are more dramatic than others—that is, they involve a larger jump in the rate of spinning. In 1993, a pair of physicists from the University of Texas at Austin studied twenty years' worth of data and found that the distribution of glitches follows a perfect, scale-invariant power law, small glitches being more common than large.[10] How might this come about? Their idea was this: Since the pure nuclear material making up a neutron star is so dense, the star's surface lives under a terrific force of gravity. Indeed, gravity is trying to crush the star down to a smaller size. You might liken this force to that applied along a fault by the movement of continents. The material of the pulsar generally resists collapsing, just as the Earth's crust resists slipping, but occasionally things give way.

In such a "starquake," the neutrons organize themselves into a somewhat smaller and denser ball, and just as an ice skater spins faster after pulling her arms in, so the star begins to spin a bit faster. If this idea is correct, then the power law for pulsar glitches simply reflects the Gutenberg-Richter law for earthquakes, as applied to the materials in a neutron star.

In the 1990s, physicists discovered the fingerprints of self-organized criticality in the crumpling of a piece of ordinary paper, the movement of magnetic fields through a superconductor, the haphazard bursting of solar flares, and even traffic jams. These examples could easily be multiplied. The critical state seems to exist in all kinds of things regardless of what they are made of, and what elements of physics go into their description. In a sense, the organization of the critical state is more basic than physics. It stands behind physics, as the ordering soul of a great deal of the world.

Rice: A Better Kind of Sand

But it is not the whole world. A tub of cold water most certainly isn't in a critical state, nor is an ordinary iron magnet or anything else under most conditions. Of course, Bak, Tang, and Weisenfeld never suggested that everything is in a critical state. They hoped that self-organized criticality might describe many things, especially things that live under conditions that are far from equilibrium. The tub of water is at rest, and in equilibrium. In contrast, the continual rain of grains falling down from above keeps the sandpile away from equilibrium. Similarly, the Earth's interior heat drives the continental plates to move ceaselessly about the surface above, keeping the Earth's crust out of equilibrium.

Even away from equilibrium, however, it is easy to find things that are not in critical states. Heat some water from below and, if the heating is strong enough, the water will break out into motion. But the motion will not be either irregular or unpredictable, and it will not show any power laws. Do the experiment carefully, as the French physicist Henri Bénard did way back in 1900, and you'll see the formation of a perfectly hexagonal array of cells (SEE FIGURE 5). Within each, water will be flowing up in the center and down at the boundaries. If some things organize into critical states, not everything does, even away from equilibrium.

So what is self-organized criticality? When does it occur? And when doesn't it? Bak, Tang, and Weisenfeld, remember, had studied a computer model of a sandpile—the sandpile game, we've been calling it—rather than a real pile. And in the early 1990s, physicists doing careful experiments noticed something that may at first seem disappointing: avalanches in real, authentic sandpiles don't quite fol-

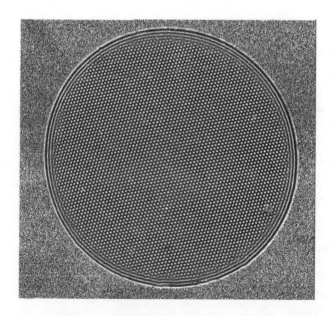

FIGURE 5. Hexagonal convection pattern in the Bénard experiment, as pictured from above. Fluid flows downward in the dark regions and upward in the light. Image courtesy of Eberhard Bodenschatz, Cornell University.

low a true power law. As it turns out, real sandpiles work less like the sandpile game than like Yellowstone National Park: they have an inherent tendency toward really large avalanches. So if self-organized criticality does apply to neutron stars, the Earth's crust, epidemics, and the way fires take place in most forests, it ironically does not apply in the setting in which the idea was first born: the sandpile. If Bak and his colleagues' computer sandpile game does indeed organize to a critical state, real sandpiles do not.

Oddly enough, this ironic twist gives the first clue to how self-organized criticality really works. In 1995, Kim Christensen and his colleagues at Imperial College in London discovered that while the sandpile game does not describe growing piles of real sand, it does describe growing piles of real grains—it is just that the grains need to be made of rice, not sand. In a careful experiment in which they dropped grains of rice one at a time into the space between two upright pieces of Plexiglas, Christensen and his colleagues found a near-perfect power law for the distribution of avalanches.[11]

They were also able to make a spectacular experimental display of the bizarre rhythm that issues from something in a critical state. They ran the experiment for a while and counted how many grains slid at each moment. They did this by taking pictures of the pile and counting the grains by hand. The results show a wild rhythm of utter unpredictability (SEE FIGURE 6). There are long periods of calm punctuated by violent outbursts of behavior—true catastrophes. For any grain of rice involved in such an event, it would be sorely tempting to wonder what made that one so big.

So why does it work for rice but not sand? The answer, it seems, has to do with inertia. Sand grains are relatively heavy, and also slippery. Once they start sliding, they have a tendency to keep going, and to bring the whole pile down. By contrast, rice grains are relatively light, and more sticky. So when the rice begins sliding, it slides only until the pile has reached the very next barely stable configuration, and then the grains stop. In the sandpile game, Bak and his col-

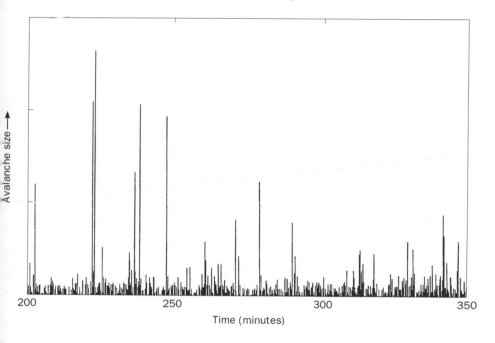

FIGURE 6. The sporadic and unpredictable record of avalanches in the rice-pile experiment. The height of each spike indicates the "size" of the avalanche triggered by the dropping of a single grain. More specifically, each spike is tall in proportion to the total loss of altitude of all the grains that participated in that avalanche, and so reflects not only how many grains slid but also how far they fell. Image from V. Frette et al., "Avalanche Dynamics in a Pile of Rice," *Nature* 379 (1996): 49–52, reprinted by permission.

leagues made their grains very sticky and gave them no inertia at all, so the grains were really much more like grains of rice. The sandpile game should have been called the rice-pile game.

Thus a simple change in the objects being dropped can mean the difference between criticality and something else. So can the rate at which grains are dropped. In 1994, the physicist Geoff Grinstein of the IBM research division pointed out that the sandpile game reaches the critical state only if the computer drops the grains very slowly. It has to drop a grain and then wait until the avalanche it triggers has come to a finish. Then it can drop another. Drop grains any faster—that is, make the next grain land before the previous avalanche has run its course—and the critical organization will go away, as will the power law.

These subtleties have a couple of important implications. Bak, Tang, and Weisenfeld had been amazed that their computer game could organize, without any tuning whatsoever, into the critical state. But it now appears that they may have tricked themselves. Grains have to be dropped very slowly, and have to be far more sticky than heavy. It appears that Bak, Tang, and Weisenfeld, in their search for novel behavior, unconsciously tuned their computer game to give it the properties it needed. Self-organized criticality seems to show up only in things that are driven very slowly away from equilibrium, and in which the actions of any individual piece are dominated by its interactions with other elements.

The belated recognition of the tuning involved in the sandpile game comes as a blow to the original dream of self-organized criticality. Bak once wrote, for example, that the notion of self-organization was of "crucial importance for the concept of self-organized criticality to have any chance of describing the real world; in fact, this is the whole idea. . . . the pile bounces back to the critical state when we try to force it away from that state."[12] If the sandpile game has to be tuned to be critical, there are scientists who can do the tuning. But how is tuning supposed to happen in the real world?

And yet this is not quite the blow it may seem to be. For one thing, almost every good theory in the world has some numbers in it that have to be tuned to make the theory fit reality. James Clerk Maxwell's equations for electricity and magnetism contain the speed of light; quantum theory contains Max Planck's constant; and applications of these theories to the real world inevitably involve these numbers as well as the masses of some particles such as the electron. The standard theory of particle physics is considered an excellent theory even though it has nineteen such "tunable" numbers. If a simple game can explain lots of things by tuning just two, that seems pretty good.

The tuning also appears to be of a rather weak kind.[13] For the sandpile game, it is only when grains are dropped much more slowly than the rate at which they topple that you find the critical state. In other words, the ratio of the rate of dropping to that of toppling has to be tuned to zero to achieve criticality. But tuning to zero, curiously, may be easier than tuning to any other number. Take some really small number such as 0.0001, change it by 10 percent, or even multiply it by 2 or 10 or 100, and you still have a very small number. This may explain why "self-organized criticality" seems to describe so many things in the real world. In earthquakes and epidemics, for whatever reason, the necessary rough tuning seems to be there.

And there seems to be no end of other examples. Even the complex process of life on Earth reveals the weird organization of the critical state, and so it also appears to be subject to unavoidable episodes of catastrophe.

· 5 ·

Killing Time

It is in the nature of a hypothesis when once a man has
conceived it, that it assimilates everything to itself,
as proper nourishment, and from the first moment of your
begetting it, it generally grows stronger
by everything you see, hear or understand.

—LAURENCE STERNE[1]

There is no such thing as philosophy-free science;
there is only science whose philosophical baggage is taken
on board without examination.

—DANIEL DENNETT[2]

✦

HELL CREEK SLIPS QUIETLY OUT OF ONE END OF THE FORT PECK
reservoir in easternmost Montana, and from there cuts a lonely,
meandering path into the hills. This is a land of vast spaces, of
prairie, grasslands, and broad valleys spotted with pine, and of weird,
twisted rocks that glow orange and purple under the blistering sun.
Hell Creek runs through those rocks, flanked on either side by
rugged outcroppings that tell the story of hot, dry summers and
freezing winters, of winds and rains slowly carving the earth into
pieces. Each year another sliver of rock becomes a whisper-thin layer
of sediment, and adds to the accumulating record of all that has happened here. For more than a century, paleontologists have been digging into that record and studying its fossils. And here they have read
the terrible story of one of the world's greatest murder mysteries.

For a paleontologist, traveling through time usually means digging into layers of sediment, but on Hell Creek it means moving
upstream toward higher strata. A morning's walk can easily cover
some 10 million years.[3] Not too far from the reservoir, the deeper
layers of sediment were put down nearly 70 million years ago when
the American plains lay under a vast, shallow ocean. These rocks
reveal a rich fossil graveyard of clams and innumerable other sea
creatures. Somewhat farther upstream the rocks are a few million
years younger; they correspond to a time when the sea had receded,
eastern Montana then being lush country of forests and rivers. Here
one finds small bits of plant material, teeth and claws, traces of the
wild and vibrant world where *Tyrannosaurus rex* fought classic battles
with its historic nemesis *Triceratops*. This is the best place in the
world to search for *Tyrannosaurus rex*; indeed, Hell Creek is still the
only place on Earth that full skeletons have been found. Just a bit
farther up the creek, however, things suddenly change. At the 65-

million-year mark, the great heel of fate seems to have come down on that fantastic world of the dinosaurs, and the sediments reflect a sudden and terrific episode of mass death. Above this KT boundary, as geologists and paleontologists call it, all traces of the dinosaurs and thousands of other species have simply vanished.

The fossil discontinuity at the KT boundary is so dramatic that geologists actually take it as defining the breaking point between two periods of geological time, the earlier Cretaceous (from the Latin *creta*, for "chalk"; in German, *Kreide*, hence the *K* in KT) and the later Tertiary. The line between the two, corresponding to a time some 65 million years ago, shows up at thousands of sites all over the world. Sites in northern Spain, for example, offer a dramatic account of the demise of the ammonites, oceanic creatures with coiled shells that were abundant in the seas for 330 million years before the KT boundary. Below the boundary, ammonite fossils are everywhere. Above it, there are none. Sixty-five million years ago, something perpetrated a colossal act of murder. In a geological blink of the eye, the dinosaurs and 75 percent of all other species suddenly went extinct.[4]

What happened? With regard to the dinosaurs, one researcher commented in 1905 that "with such an excessive load of bony weight entailing a drain on vitality, it is little wonder that the family was short lived."[5] Another questioned the dinosaurs' sexual drive, noting that "the increase of weight in the limbs of the dinosaurs recalls the eunuch condition."[6] At one time or another, the dinosaurs' extinction has been attributed to their going blind, to greedy mammals eating all their eggs, to volcanoes blasting them into oblivion, or to the effects of a climate that suddenly became too cold, too hot, too dry, or too wet.[7] Of course, it was not only the dinosaurs that disappeared.

At scientific meetings and in scholarly journals the world over, heated arguments continue about what it was that nearly wiped life off the planet. Scarcely a week goes by without someone publishing "evidence" for this or that scenario, or "evidence" against another.[8]

But if no one can say for sure what happened, there is one thing on which all scientists can agree: whatever the cause, it was not so terribly unusual. A closer look at the fossil record shows that devastating catastrophes have struck the Earth many times, and the KT extinction was not even the worst.

Well below the KT boundary, another dramatic line of death cuts through the rock, this one laid down 250 million years ago. Again, the break in the record of life is so clean that geologists take it to define the end of the Permian period of geological time and the beginning of the Triassic period. In 1998, the geologist Samuel Bowring of the Massachusetts Institute of Technology and some colleagues dated the duration of this episode of extinction and found that it took place within a time span as short as ten thousand years. That may seem like a long time, and yet, in a history of multicellular life spanning some 600 million years, it amounts to a mere instant. "The event," Bowring and his colleagues wrote, "marks the most widespread annihilation of life in the past 540 million years." A full 95 percent of all species in the oceans were killed, as were a comparable fraction on land.[9]

Similar catastrophes struck the planet 440, 365, and 210 million years ago. Together with the KT and Permian disasters, these mass extinctions form a Big Five that stand out in the record of all extinctions much as large earthquakes in 1836, 1838, 1868, 1906, and 1989 dominate the seismological history of the area around San Francisco. Charles Darwin once wrote that in the light of the theory of evolution, "the old notion of the inhabitants of the Earth having been swept away at successive periods by catastrophes, is very generally given up."[10] Evidently not. If the fabric of life appears to be stable and slowly changing, the historical record says otherwise. And the Big Five extinctions are merely the most obvious of the sudden disruptions. In between, the record shows countless smaller mass extinctions. Life on Earth seems to suffer sporadic and catastrophic episodes of collapse.

But not *every* extinction is part of a *mass* extinction. Evolutionary biologists estimate that some few billion different species have evolved at one time or another during the course of life's history. Only a few tens of millions exist today, however, which means that 99 percent of all species in history are now extinct. Extinction is so natural an event in evolution that, as someone once said, "To a first approximation, everything is extinct." As it turns out, only 35 percent of all species died out as part of a mass extinction. The "background" extinctions account for nearly two-thirds of all extinctions. But they do not account for the great waves of mass killing. What could lie behind this strange record of sporadic catastrophe?

Acts of God

The insurance industry uses the phrase "acts of God" to refer to accidents or disasters that are simply beyond the power of anyone to foresee, and for which no one can sensibly be blamed. If a tornado plays Frisbee with the roof of your house or a bolt of lightning torches your brand-new Porsche, you are the victim of an "act of God." The forces of destruction have paid you an unfortunate visit, and the insurer pays—unless, of course, your insurance contract specifically excludes such accidents from coverage, in which case it is just your tough luck.

In the mass extinctions, as opposed to the run-of-the-mill background events, most scientists see the consequences of similar "acts of God." We all depend for our continued existence on a stable and accommodating environment. We need oxygen, the right temperature, plentiful water and food, not too much radiation, and so on. If something goes terribly wrong with the environment, life suffers. For most scientists, then, explaining the mass extinctions means finding out what went wrong with the environment, and why. Things can go wrong in a thousand ways, of course, but two kinds of happenings have come to be seen as especially significant. One is rather more violent than the other.

In 1980, a team of scientists led by physicist Luis Alvarez of the University of California at Berkeley proposed that the KT disaster was the direct result of a worldwide atmospheric upheaval triggered by the terrific impact of a huge asteroid or comet on the Earth.[11] "Probably ten kilometers across," one of the researchers has written, "and travelling tens of kilometers a second, its energy of motion had the destructive capability of . . . ten thousand times the entire nuclear arsenal of the world."[12] The impact would have vaporized rocks, blasted a hole in the ground 40 kilometers deep, and launched through the atmosphere and into space an immense volume of tiny rock particles and ultrafine dust. The tiny particles, falling back to Earth, would have scorched the air within a thousand miles of the impact site, "cooking, charring, igniting, immolating all trees and all animals which were not sheltered under rocks or in holes. . . . Entire forests were ignited and continent-sized wildfires swept across the land."[13] But this only set the stage for the real killer. The dust, scattered throughout the upper atmosphere, would have blotted out the Sun for months, producing continuing night. Plants would have withered. Herbivorous animals would have died of starvation. Devastation would have rippled up the food chain and brought it crashing down like a house of cards, eventually wiping out even the most ferocious carnivores.

If this idea seems like science fiction, it is supported by a good deal of evidence. To begin with, scientists have found significant quantities of the rare element iridium in the rocks at the KT boundary, not just in one place, but at more than a hundred sites worldwide. Just after the Earth formed, when it was a hot molten blob, the heavier elements sank toward the center. As one such element, iridium is only rarely found in the crust. So how did it get into the KT layer? Well, asteroids and comets contain quite a lot of iridium. Whatever it was that blasted the planet, its iridium seems to have found its way into the upper atmosphere, filtered around the globe, and then settled back into the KT layer. Scientists have also measured the levels of other rare elements in the KT boundary, such as ruthenium and rhodium, and the ratio of their abundance is just as it is in asteroids and comets.

If that is not convincing, eleven years after Alvarez and his colleagues suggested their impact scenario, another team of scientists discovered an enormous crater in the Yucatán Peninsula in Mexico.[14] You could stand on top of the Chicxulub crater, as it is named, and not know anything was there, for it is buried about a mile beneath the surface of the Earth. Yet the crater is nearly 180 kilometers across, and in 1992, when it was possible to establish its age, the crater turned out to have been made 65 million years ago.

Not that there aren't problems with this idea. The very fact that there is a crater implies that there was a huge impact. The iridium deposits and other evidence indicate that the physical consequences of the impact stretched over the whole world. But was that enough to trigger a mass extinction? It has been pointed out that the bulk of the long 1980 paper by Alvarez and his colleagues was "confined to the geological and physical evidence for an impact, and the physical results of the impact." The discussion of the biological results of the impact occupies only half a page.[15] The reason is simple. No one really has much of a clue about what an impact would really do to life all over the planet.

What is more, it is puzzling that some species died out while others came through unscathed. As one of the Alvarez team admits: "Many smaller land animals survived, including mammals, as well as reptiles such as crocodiles and turtles. No one really understands why these animals escaped extinction."[16] To make matters even more puzzling, other tremendous impacts in the past haven't seemed to harm anything. In 1998, the geologist Ken Farley of the California Institute of Technology and other researchers uncovered evidence that a huge crater 100 kilometers across in northern Siberia was created at precisely the same time as an 85-kilometer hole at the mouth of Chesapeake Bay in the United States. Both were made 35 million years ago when comets smashed into the Earth during a comet shower in the solar system. The fossil record shows absolutely nothing unusual at the time.[17]

In view of these outstanding questions, not everyone believes that the dinosaurs were wiped out by a fatal rock from the sky. Scientists are kicking around a few other ideas as well. The dinosaurs, of course, were not living on Earth all by themselves. Some years ago the geologist Leigh Van Valen of the University of Chicago noted that the mammals began thriving and increasing their numbers just a few hundred thousand years before the KT mark, and that they could have muscled the dinosaurs out of existence. Some paleontologists suggest that the battle may have been swayed also by changing climate, as the dinosaurs' lush, warm habitats were then turning into cool forests more suitable for the mammals. This same climate change may also have triggered the extinctions of many other species. Perhaps the impact had nothing to do with the KT extinctions at all.

Chill Winds

What of the other mass extinctions, 210, 250, 365, and 440 million years ago? For these events, no one has yet found a huge crater of just the right age. They may do so in the future, but for now most paleontologists suspect that something else was at work, such as sudden changes in climate. According to Steven Stanley of Johns Hopkins University, "There is one simple fact that makes climate change a likely general cause for mass extinctions. This is the relative ease with which change in global temperatures can eliminate myriads of species."[18]

Every species, after all, is adapted to a certain climate, whatever it might be. If temperatures fall, for example, the species needs to migrate toward the equator to maintain its climate, or adapt to the new, colder conditions. But some species can't migrate, being blocked by mountains or large lakes or oceans, or because they live in the treetops of forests that don't extend farther south. Likewise, if the temperature drops too quickly, a species may not be able to adapt rapidly enough, and will simply go extinct.

Global temperatures did drop precipitously in the case of the Permian extinction, the biggest of them all, 250 million years ago. Some other ominous things happened then too. For one, the level of the seas dropped markedly. When sea level falls, the oceans recede from the continents, exposing vast areas of continental shelf to the atmosphere. These shelves contain enormous amounts of organic material, which, when it interacts chemically with the atmosphere, would take up vast quantities of oxygen. The paleontologist Paul Wignall of Leeds University in the United Kindom has estimated that this interaction may have brought oxygen levels down to only half of what they are today. "The Permian-Triassic mass extinction," he concludes, "appears to be a story of death by suffocation."[19]

There is no definite consensus yet on which of these influences were more important, or on whether they were also possibly at work in the other major extinctions. Some paleontologists point to dramatic periods of volcanic activity that spewed untold volumes of dust into the air. Others see the workings of extended global droughts behind mass extinctions. A list of all the proposed causes would go on for pages, and well might make you wonder how closely any of these ideas are tied down by the facts. After all, it is a virtual certainty that, whether 65, 210, or 250 million years ago, something unusual was happening on Earth—temperatures or sea levels were rising or falling, volcanoes were erupting, the amount of ultraviolet radiation from the Sun was increasing, and so on. The consequences of any of these things for the global ecosystem remain largely a matter of guesswork. So it is, perhaps, little wonder that so many potential causes have been considered. As paleontologist David Raup has wondered, "Could it be that the list of probable causes of extinctions is simply a list of the things that threaten us as individuals?"[20]

Nevertheless, almost all scientists do agree that mass extinctions are caused by some combination of shocks or changes that disrupt the environment in which organisms live. This, it seems, must be what makes the mass extinctions stand out from the ordinary background extinctions that go on all the time. During the normal periods, the ordinary workings of evolution hold sway, condemning now one, now another species to oblivion. But these periods are interrupted by intrusions from the outside. "The alteration of background and extinction regimes," the paleobiologist David Jablonski has put it, "shapes the large-scale evolutionary patterns in the history of life."[21] Or, as Richard Leakey and Roger Lewin have described the pattern of change in the world's ecosystems:

> That pattern includes two phases, the intervals of background extinctions, in which species disappear at a low rate, and episodes of high rates of extinctions, including major biotic crises. Most biologists agree that the prevailing force in times of background extinctions is natural selection, in which competition plays an important part.[22]

Most biologists do not believe that evolution is itself capable of causing the terrific upheavals. There is evolution, and there are the upheavals caused by shocks from the outside. This is a simple and satisfying picture, and yet it appears to suffer from a serious problem.

Ten Years in the Library

Jack Sepkoski of the University of Chicago is a paleontologist who prefers to do his research not in the field but in the library. He doesn't dig for fossils. He digs for information about fossils that others have previously discovered. As a graduate student at Harvard in the mid-1970s, Sepkoski began stuffing notebooks full of data cobbled together from books and research papers, and from conversations with friends and other paleontologists. His aim? To compile an enormous catalog showing when groups of organisms came into being, and when they went extinct.

In thinking about evolution, most people tend to think about species. Sepkoski didn't. The tree of life has thick limbs that give way to thinner ones, off of which sprout smaller branches, and branches off those branches, down to the smallest discernible twigs—the species. Quite sensibly, biologists have words for all these various categories. A genus is a collection of several closely related species, and a family is a group of related genera (plural of "genus"). In 1982, Sepkoski published the first installment of his version of the fossil record—a massive database documenting the origination and extinction of many thousands of families. That was only a stop along the way. After another ten years of gathering, he had assembled a database for some forty thousand different genera (all marine invertebrates) falling into some five thousand different families (with about eight genera in each family).

Sepkoski chose the marine invertebrates—things such as sponges or corals—since these form the largest part of the fossil record. And when he finished in 1993, researchers found themselves with a condensed picture of how life has flourished, and perished, on this planet over the past 600 million years.[23] Sepkoski's efforts stimulated further work, and soon after, geologist Michael Benton of the University of Bristol in England finished compiling an independent database documenting the times of origination and extinction of some seven thousand families of organisms, in this case, both marine and terrestrial varieties.[24]

Before we see what this wealth of data can reveal, it is worth saying a bit about just how difficult it is to assemble. "Ten Years in the Library," the title of Sepkoski's 1993 article in *Paleobiology*, gives some idea of the immense size of the literature, but it doesn't hint at the tricky problems that researchers face in doing this sort of thing.[25] One problem is the "pull of the recent." Only a tiny handful of the organisms alive at any time end up as fossils. Most simply die and their bodies decay and that's that. But of those that do form fossils, the ones created in recent history have a better chance of surviving intact to the present. So the fossil record is inevitably skewed toward the present.

Then there is the "monograph effect." If very few fossils persist into the present, then those that researchers actually discover are far fewer still. It is certain that many species remain undiscovered. As a result, if one energetic researcher does an unusually thorough job on a particular period of the geological past, the fossil record ends up showing a blip there, as if there had been a sudden burst in the number of species. There are many other problems too. One particularly annoying uncertainty lies behind Sepkoski's and Benton's decisions to study extinctions at the level of genus and family rather than species.

To learn as much as possible about the comings and goings of living things, it would clearly be best to work at the level of species. But the fossil record is sparse, and that makes for trouble. Suppose you have dated some fossils of a species and you are trying to estimate when that species went extinct. As an estimate, you might take the date of the youngest of all your fossils. If you have only a few fossils, however, your estimate will not be very good. In 1982, the geologists Phil Signor and Jere Lipps of the University of California at Davis pointed out that if you have only a few fossils, it is unlikely that any one will fall very close to the time of extinction of the species. It is far more likely to fall somewhere roughly in the middle between its beginning and end. Consequently, you will estimate that the species went extinct earlier than it did, and the fewer fossils you have, the greater the error. This Signor-Lipps effect, as it's known, also has a converse, known jokingly as the Lipps-Signor effect, which affects estimates of the origination times of species. Taking the date of your oldest fossil as a guide will make it seem as if the species originated after it really did.

All told, the effect of a sparse fossil record is to make species seem to originate later and die out sooner than they really did. Fortunately, the errors grow smaller with the presence of more fossils. And that is why Sepkoski, and later Benton, decided not to study species, but to look higher up in the tree of life at the level of genera or families. Working here, they could pool the fossils of different species and obtain more accurate estimates of when any particular group originated or disappeared. In assembling the data, they took pains to correct for the pull of the recent, the monograph effect, and other known biases, much as historians take documents not just at face value, but with a healthy dose of skepticism. As a result, Sepkoski's and Benton's data represent the best picture we have of the true pattern of past life on Earth.

As we have seen, the orthodox view on extinction holds that there are two kinds: background extinctions, caused by ordinarily evolutionary processes, and mass extinctions, triggered by climatic changes, asteroid impacts, or other shocks to the biosphere. A rough plot of Sepkoski's and Benton's data seems to back up this point of view. The record of the fraction of families going extinct in each geological period shows a pattern of relative quiet punctuated by sudden cataclysms (SEE FIGURE 7). The tremendous extinctions stand out from the rest. But this is a story we have heard before. Plotting how often extinctions of various sizes happen tells a very different story.

In 1996, the physicists Ricard Solé and Susanna Manrubia took a more careful look at Sepkoski's data and found that the distribution of extinctions according to their size (this being taken as the number of families that went extinct) follows our old friend the power law. In fact, the pattern of regularity is in this case identical to that for earthquakes: if you double the size of the extinction under consideration, you find that such events become four times as rare. And this power-

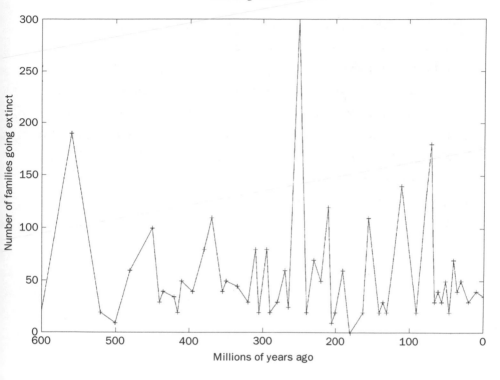

FIGURE 7. The record of the mass extinctions. Each peak corresponds to the number of families that went extinct in each of the various geological stages. The five largest peaks correspond to the greatest mass extinctions, the last being the one that wiped out the dinosaurs 65 million years ago.

law regularity extends from extinctions involving just a few families all the way up to the worst events that wiped out thousands.

Do violent happenings generally imply violent causes? Must every dramatic extinction have an equally dramatic cause? We have seen in earlier chapters that this prejudice has taken a beating in recent years—for example, in the context of earthquakes and forest fires. The remarkably simple form of the curve for mass extinctions hints that scientists may be making a terrific mistake in thinking of these "standout" episodes as something special. It is no doubt true that if you place spikes on a time line showing when the various extinctions happened, and how big they were, the biggest spikes stand out to the eye as something "obviously" special. But displayed in a different way, the same record seems to show that there really is nothing remarkable about the biggest events. The power-law perspective hints that the mass extinctions may not be exceptions to the workings of evolution. Rather than the fingerprints of the Hand of God reaching in from afar, they may be the inevitable product of evolution's most ordinary principles.

· 6 ·

The Web of Life

A cultivated person's first duty is to be
always prepared to rewrite the encyclopedia.

—UMBERTO ECO[1]

Things should be made as simple as possible,
but not simpler.

—ALBERT EINSTEIN

IN THE LATE 1970S, A MINOR ECOLOGICAL CATASTROPHE WAS BREWING in the grassy countryside of southern England. Hordes of rabbits were devastating hundreds of thousands of acres of rich farmland. Fortunately, the British government had a safe and easy biological solution ready to hand. The myxomatosis virus thrives almost exclusively in the bodies of rabbits. It does not kill them, but makes infected animals sluggish, thereby slowing their breeding and making them more susceptible to predators. By introducing myxomatosis, authorities reasoned, they could manage the rabbit population with little adverse effect on the balance of the countryside ecology. Things were not that simple.

Myxomatosis did bring the number of rabbits crashing down within a few years. Meanwhile, however, livestock prices fell, and grazing animals on grasslands became relatively unattractive to farmers. With fewer animals grazing, and fewer rabbits nibbling, the grass in the fields of southern England grew taller than usual. This doesn't sound particularly grave. But there is an ant named *Myrmica sabuleti* that thrives in short grass, and does less well in longer grass, and soon grassland populations of *Myrmica* were decimated. This ant has a peculiar relationship with the large blue butterfly *Maculinea arion*. When this butterfly lays its eggs, the ants carry them into their burrows and foster the larvae through hatching and into adulthood. Unfortunately, the population of *Maculinea arion* was already struggling in the late 1970s, and when the number of ants fell, the number of butterflies plummeted. The introduction of myxomatosis made for taller grass and fewer ants and obliterated from England a beautiful blue butterfly.

Unpredictable ecological chain reactions such as this are hardly unusual. To probe the structure of an ecosystem, ecologists often do experiments in which they remove one particular kind of predator, and then look for the effects this has on its principal prey. You might expect the results to be fairly predictable: remove the predator, and the numbers of its prey should increase. But in 1988, Peter Yodzis of the University of Guelph, in Canada, found otherwise.[2] In a study of thirteen different ecosystems, he found that the number and complexity of the indirect pathways linking species was so great that the effect of removing a predator was largely unpredictable—even the effect on its obvious prey. Removing a bird that eats a certain kind of mouse, for example, might ultimately make for fewer mice rather than more. If the bird has an even greater appetite for some of the mouse's direct competitors, its absence will boost the competitors more than the mouse, and the population of mice will suffer. By such indirect pathways, a change in the numbers of one organism might have all kinds of unforeseen consequences, even for species that seem completely unrelated.[3]

A more recent and more comprehensive study leads to a similar conclusion. Every year, the North American breeding bird survey estimates the populations of more than six hundred different bird species in the United States and Canada. The survey has been carried out for more than thirty years, yielding a wealth of detailed information about the fluctuations in numbers of all these species. In 1998, the ecologist Timothy Keitt of the University of California at Santa Barbara and the physicist Gene Stanley of Boston University used the database to calculate, for each species in each year, how fast its population was changing. A population might go up by 10 percent one year, and fall by 15 percent the next. Looking at such changes, Keitt and Stanley did for population growth rates what Gutenberg and Richter did for earthquake intensities. That is, they looked to the record and worked out how often they found one rate of change relative to another.

So what is the typical rate of growth (or decline) for bird populations? Remarkably, there isn't one. If the most likely change is no change at all, changes either upward or downward are progressively less likely, and both follow an identical power law. "For the species considered here," Keitt and Stanley concluded, "there is no characteristic scale of fluctuation in population size. . . ." In other words, what will happen next is unpredictable not only in its direction, up or down, but even in the rough scale of its magnitude. This is, of course, exactly the same kind of scale-invariance we have seen in the critical state. If ecosystems live in a critical state, then large upheavals should be expected, and we should expect to find scale-free distributions everywhere. So you might well wonder: Does this have anything to do with the power law for mass extinctions we met in the last chapter? Could the mass extinctions arise solely from the internal workings of the ecosystem? It is an intriguing idea.

Keitt's and Stanley's data, however, refer to ecological chains of cause and effect. These involve populations of different species interacting with one another, and can lead to marked changes in the number of organisms in just a generation or two. Biologists contrast such fast ecological dynamics with evolutionary changes that take place far more slowly, and typically become visible only over many generations. Evolution changes not only numbers but also the character of organisms—leading to longer beaks, brighter feathers, or spots on a tail fin. Since the fossil record extends over a terrifically huge stretch of time, any explanation of the mass extinctions in terms of ordinary internal workings of an ecosystem should really look to patterns in evolution, rather than ecology.

Of course, evolution is fully capable of driving species to extinction. As conditions change, a species may fail to adapt. It is easy enough to imagine, then, that since no species lives in isolation, one such extinction might trigger another, which could in turn trigger yet another, setting off a deadly avalanche that might well propagate a long way. Is there any evidence that runaway extinctions of this sort have something to do with the mass extinctions? The answer, as we shall see, is yes. Many scientists have come to suspect that the global ecosystem is tuned to a critical state not only in its ecological but also in its evolutionary workings, and that the extinction of just one species can sometimes trigger a system-wide catastrophe.

Wandering in the Hills

Stuart Kauffman is a physician by training, thinks more like a physicist, and works almost exclusively on problems in biology—not small problems, such as pinning down the structure of one more protein or the sequence of one more gene, but large problems. In the mid-1980s, while at the University of Pennsylvania in Philadelphia, Kauffman proposed a radically new view of the origins of life on Earth. Before then, all theoretical estimates made it seem that the origin of life had been a stupendously improbable event. In the primordial soup several billion years ago, some first bundle of molecules had somehow arisen that could reproduce itself. Once the evolutionary mechanism of "differential survival of replicating entities" got going, of course, it could keep going. But how did it start in the first place? Estimates inevitably showed that there had not been enough time since the beginning of the universe.

Kauffman began playing with simple mathematical games designed to mimic the interaction of molecules of different kinds on the early Earth, and he made an intriguing discovery. Some molecules act as catalysts—that is, they act to speed up the chemical interactions between others. Molecules of type A, for example, might greatly enhance the rate at which types B and C can come together to create type D. Kauffman studied random networks of interacting molecules. If the number of kinds of such molecules was small, he found nothing exceptional in the chemistry of the soup. But as he increased that number, he found that there would inevitably arise in the network what he called an "autocatalytic set." This is a subset of molecules that can pull themselves up by their bootstraps. Molecule

A might help to catalyze the production of D; D might help to produce E or F, which would in turn catalyze C and G, and so on. Ultimately, far down the line, Y and Z might catalyze A and B and complete the linking up, so that every molecule would be catalyzed by some other one.

If such a positive feedback loop should ever arise in a collection of molecules, the concentrations of all members of the autocatalytic set would take off. Kauffman's astonishing discovery was that the existence of such loops was absolutely certain if the number of kinds of molecules in the soup was large enough. And this number did not have to be very large. A molecular mixture shows a natural transition from boring to fascinating behavior as the number of kinds of molecules grows. This transition forms the basis of a completely new theory in which the origin of life is not improbable, but inevitable.

Inspired by the power of this way of thinking, Kauffman turned his attention from the origin of life to its rhythms once established— that is, to evolution within the complex setting of ecosystems. This problem, of course, is horribly complicated, because the environment in which a species evolves is hardly ever fixed. Species interact, they eat one another, compete for territory, form cooperative habits, and so on, and when one species evolves, this alters the evolutionary conditions for others. To try to gain insight into such problems, evolutionary biologists spend much time thinking about landscapes— not real landscapes, but undulating mathematical surfaces known as fitness landscapes. These surfaces enter directly or indirectly into almost every powerful argument that has ever been made about the nature of evolution, and Kauffman's is no exception.

Evolution accomplishes its ends through the triple action of variation, selection, and replication. In any population of rabbits, for example, some will see better, run faster, or think quicker than others. This is variation. These fitter rabbits tend to live longer and produce more offspring than do weaker rabbits. This is selection. And because parents pass copies of their genes down to their offspring, the next generation almost certainly contains a greater proportion of fitter rabbits than did the last—a consequence of replication. As a result, the overall fitness of the population slowly increases.

The notion of a fitness landscape helps us to visualize such changes more precisely. In biological vernacular, an organism's color, speed, strength, intelligence, and so on make up its phenotype. Phenotype determines fitness. Now imagine a two-dimensional grid, with each point corresponding to a different phenotype, and picture an undulating landscape running above this grid, the height at every point being the fitness of an individual of that phenotype (SEE FIGURE 8). The landscape rises up and down as the phenotype changes. This is a fitness landscape.

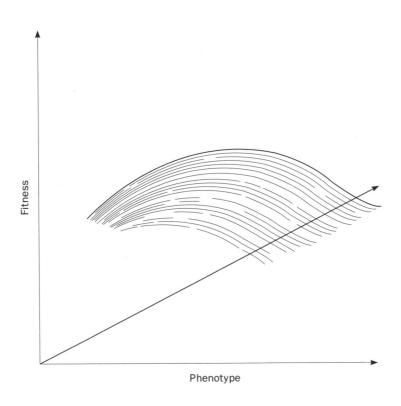

FIGURE 8. A fitness landscape with a single peak.

Viewed on such a landscape, evolution is little more than populations climbing into the hills. Some misfit rabbits might start out as
a cloud of dots down in a fitness valley. Some will get eaten, some
will reproduce, and after one generation the cloud will make a new
cloud. Generation after generation, cloud will replace cloud, and
each time, because of the shuffling of genes and random mutations,
the cloud will kick a few dots to places just a bit higher than before.
Since these fitter individuals generate more offspring, the cloud will
drift uphill until it reaches a peak.

I have been speaking so far as if the shape of the fitness landscape
depends only on the organism in question. But it also depends on
lots of other things, including the climatic conditions and the characteristics of all the other species of trees, birds, and bacteria with
which the organism interacts. A population of frogs may inhabit a
local peak on its fitness landscape. But if a snake that eats frogs suddenly comes into the area, the frogs' fitness landscape will shift, and
they may find themselves plunged into a deep fitness valley. Many
will end as snake food. But the population may well evolve, develop
camouflage, and make its way to another local fitness peak.

So as a result of the interaction between species, any evolutionary change in one can trigger evolutionary changes in another. In
view of this theoretical possibility, Kauffman wondered if such
coevolution might lead to some interesting effects. Unfortunately,
ecosystems are so intricately interlinked that no one has much idea
of the actual shapes of the fitness landscapes for the species in any
real ecosystem. But we have seen already how scientists have gained
deep insight into earthquakes, forest fires, and other complex processes by virtue of extremely oversimplified models, or games. In the
case of coevolution, Kauffman, working with his colleague Sonke
Johnsen, decided to follow a similar line of attack.

Digital Species

To keep things simple, they represented species as strings of 0s and 1s, and programmed a computer to keep track of the fitness landscapes for a collection of such species, as well as the species' positions on the landscapes. For each species, the landscape was chosen to reflect some of the typical features of real fitness landscapes—in particular, the presence of numerous peaks and valleys. In the simplest way possible, Kauffman and Johnsen also let the species interact, so an evolutionary change in one could affect the fitness landscape for another. They set things up with the core logic of coevolution in place, even if most of the real details were missing. Then they set the computer running.

At first, their ecosystem did nothing very interesting. All the species evolved until they lived in fitness peaks, and that was it; nothing ever changed again. The game was boring. But as they grew more familiar with it, Kauffman and Johnsen learned that by tuning their ecosystem—especially how rugged the landscapes were—they could make its evolution become far more active and vigorous. When the landscapes for each species were rugged, but not too rugged, and when the influence of one species on the landscape of another was just right, they found that their ecosystem worked like the sandpile game. An evolutionary change in a single species could trigger an avalanche of coevolution that could affect anything from a few species to almost every single species in the ecosystem.[4]

Indeed, in running their game for a long time, they found that the distribution of avalanches according to the number of species involved followed a power-law form. In the Kauffman-Johnsen ecosystem, then, at least when it is properly tuned, there is no typical size for events. When one species evolves, it may be an isolated event, or it may trigger a million others to follow suit. What is more, the record of evolutionary events in this ecosystem game looked much like the record of the mass extinctions, with long intervals of apparent quiet punctuated by sudden bursts of activity. Kauffman and Johnsen had tuned their ecosystem to the critical point, just as Fermi had his reactor.

This game, to be sure, offers a vastly oversimplified view of coevolution in the real biological world. Even so, the litany of examples we have seen so far suggests that details often have little effect anyway. In 1993, as if to prove this point, Per Bak and Kim Sneppen, both working then at the Niels Bohr Institute in Copenhagen, found an even simpler game for coevolution that leads to almost exactly the same results. It is worth looking at their game in a little detail, for it is probably the most basic skeleton of coevolution conceivable. To see where it comes from, we need to think a bit more about the nature of evolution of fitness landscapes. Climbing uphill is just part of the story.

Most real landscapes have some degree of ruggedness, with peaks of various sizes interspersed with valleys, and fitness landscapes are no different. As a result, it is unlikely that a population climbing into the hills will find its way to the highest of all peaks. Instead, it will almost certainly ascend toward one of the myriad smaller peaks, and then get stuck on top, as would any climber trying to ascend a rugged mountain without a map. Once there, a species cannot climb any higher without making a leap across to the closest higher peak. How long will it take a species to make such a leap?

To do so, some members of the population need to traverse the intervening Death Valley of lower fitness. In each generation, the cloud will throw a few new variants into the valley. Being of lower fitness, the families of these variants will generally survive for only a generation or two. On rare occasions, however, a family might last four or seven, or even ten, generations. Eventually, on one of these occasions, an unspeakably rare string of genetic events, one in each generation, may lead a sequence of variants across the valley, and land a descendant on the Promised Land of the other nearby hill. Multiplying and climbing, this descendant and its offspring will then

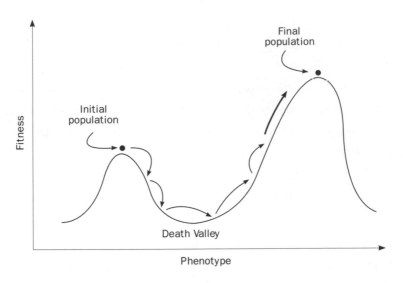

FIGURE 9. A population hung up on a fitness peak can make it over to another only by mutating its way through an intervening valley of lower fitness.

carry a population upward to the higher peak (SEE FIGURE 9). Evo-lutionary theory predicts that the time required to make such a jump becomes very large very quickly as the distance to the next peak grows.[5]

In other words, if a species faces a short jump, it might make it in a reasonable time. If it faces a long one, then forget it—it will be hung up on that peak for eons. Bak and Sneppen used this insight to great effect. They reasoned that each species, after climbing into the hills, will inhabit a local peak and be prevented from moving to a higher peak by some intervening gorge. The width of the gorge reflects just how difficult it is for the species to evolve further, and how long it will take to do so. The widths of these gorges—one for each species—are crucial, since nothing can happen until some species makes a leap. So Bak and Sneppen focused entirely on these numbers, and ignored everything else.

Sticks and Wedges

To visualize the state of an ecosystem through their eyes, imagine a string of sticks with lengths equal to the width of the various gorges the species are facing. Just to butcher reality a bit further, Bak and Sneppen supposed that the lengths of these sticks can only be between 0 and 1. The ecosystem evolves by two rules. Since short leaps are so overwhelmingly more probable than long leaps, the first species to jump will almost always be the one facing the narrowest gorge. When this species does jump, it will find itself on a new peak, and so face a new gorge, which will be either broad or narrow—no one can tell. So, rule number one: Find the species with the shortest stick, and replace it with another having random length between 0 and 1. This rule captures how species evolve by themselves (SEE FIGURE 10).

The essence of coevolution is that species interact. For simplicity, Bak and Sneppen supposed that each species interacts with its two nearest neighbors. When one species evolves, it disturbs the fitness landscapes of its neighbors. Having been on peaks, they suddenly find themselves off peaks, and so rapidly evolve to reach new peaks, where they face new gorges. Hence, rule number two: After you have replaced the shortest stick with another of random length, also replace the sticks to its left and right with new sticks of random length. To play the game, you begin with a random assortment of short and long sticks, representing an ecosystem in some state. You repeat the procedure again and again. What happens?

At first, there are plenty of short sticks. But since the shortest stick and two others get replaced each time, the average length of the sticks increases. Eventually, as it turns out, all the sticks come to have lengths two-thirds or greater. At this point, the ecosystem has reached a relatively steady state, with all the species facing wide gorges. The ecosystem has now to wait a very long time for the next

(a)

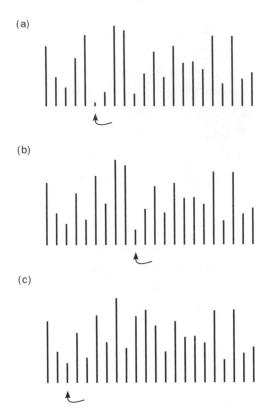

(b)

(c)

FIGURE 10. Rules of the Bak-Sneppen game. Starting from the initial state (*a*), the shortest stick (marked by the arrow) and its two neighbors are replaced by new sticks of random lengths. Each successive step in the evolution (*b, c*) replaces the shortest sticks with new sticks that tend to be longer, and so the length of all the sticks gradually increases.

evolutionary jump. Somewhere in the line, however, there is a shortest stick. Wait long enough, and the species corresponding to this stick will jump to a new peak.

This single evolutionary jump replaces the stick for one species and its two neighbors by new random sticks. All three sticks might in fact be fairly long, in which case the ecosystem will again be locked

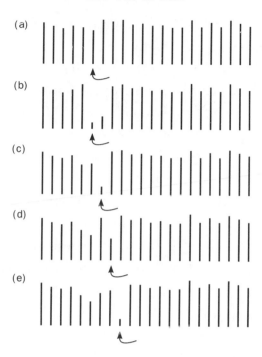

FIGURE 11. Avalanches in the Bak-Sneppen game. After things settle down and all the sticks have grown to lengths of about two-thirds or greater, the ecosystem is poised in a critical state. Now the replacement of the shortest stick has a good chance of triggering an avalanche of subsequent replacements that travels a long way through the ecosystem.

into a long-lasting situation. But there is a good chance that one of the three new sticks will be much shorter than two-thirds (SEE FIGURE 11). If so, then the species corresponding to this stick, facing a narrow gorge, will leap again very quickly. So you replace this stick and its neighbors with new sticks. Once again, one of the new sticks will probably be quite small, and this will lead very quickly to a further evolutionary jump. In this way, an avalanche of evolution will run through the ecosystem until finally, by chance, all the sticks once again have lengths greater than two-thirds. At this point, the avalanche has stopped, and with all species again facing wide gorges, there is another long wait until the next burst of activity.

So the Bak-Sneppen ecosystem evolves into a state in which all the species face fairly large barriers to further evolution. At the same time, even a single evolutionary step made by one species can destabilize the situations of other species, and can trigger rapid chain reactions of evolution that sweep through most of the ecosystem before things again settle down. Although obscured by more details, the Kauffman-Johnsen game has essentially the same character. Both suggest that the ordinary evolutionary workings of ecosystems should lead inevitably to dramatic upheavals having no identifiable cause whatsoever. Could this be the real cause of the mass extinctions?

Not surprisingly, biologists have criticized both games for all kinds of reasons.[6] These games are indeed terrific oversimplifications of biological reality, and both have some very real shortcomings. If you look at the preceding discussion, for example, you will find that neither model ever mentions the word "extinction." Indeed, they are models for coevolution in the absence of extinction. When a species hops from one peak to another, it does not go extinct; it simply changes phenotype. The avalanches are simply avalanches of evolutionary activity.

One might well argue, however, that if a hundred species are forced to adapt to new fitness peaks during some evolutionary upheaval, some will not make it and will instead go extinct. If the upheaval involved a thousand or ten thousand species, a correspondingly larger number would go extinct. This would lead to a power law for extinctions, as well as for evolutionary avalanches. This is hardly a watertight argument, but it is certainly plausible. All in all, these games suggest that the global ecosystem rests in a critical state, and they hint—but only hint—that the mass extinctions may simply be the rare but expected and natural result of ordinary evolution.

Then again, both of these games were only intended to take a first stab at the problem.[7]

Evolutionary Thinking

As it turns out, the Kauffman-Johnsen and Bak-Sneppen games probably do dispense with just a bit too much detail. If they capture the self-similarity of evolution through the power-law form for coevolutionary avalanches—a terrific achievement in itself, to be sure—the precise numbers that appear in the power laws do not quite fit reality. In the Bak-Sneppen game, for example, extinctions become about 2.14 times less likely each time the size is doubled, whereas the fossil record indicates that they really become 4 times less likely. The precise numbers in the Kauffman-Johnsen game suffer similarly. So if these games capture the existence of a critical state, and the extreme sensitivity of the ecosystem, they do not get the details quite correct.

For now, there is no consensus among scientists about which details need to be added back in, but one notable possibility has emerged from the work of the physicists Luis Amaral of the Massachusetts Institute of Technology and Martin Meyer of Boston University. An odd feature of the Bak-Sneppen ecosystem is that all species stand on an equal footing; there aren't any predators or prey. Real ecosystems, of course, have levels: there are food chains, with some species at the top, and others near the bottom. In 1999, Amaral and Meyer invented a simple game that pays more attention to the existence of food chains. Their model is really not much more complicated than that of Bak and Sneppen, and yet it gets the precise numerical form of the power law exactly right—indicating that the existence of food chains may be an important detail.

In their game, an ecosystem had six levels, ranging from the highest predators down to the lowest prey. On each level, there are a thousand ecological niches—that is, positions that can be occupied by some species. Each species from one level is assumed to feed on several from the level below. Amaral and Meyer started out with the lowest level partially filled with species, and they let things evolve according to a few simple rules. At every time step, there is a small chance that each existing species will create a new species, which will then randomly occupy any open niche on the same level, or on the one above or below. This lets the low-level species eventually spread and fill out the food chain. Each new species is automatically assigned a few species from the level below on which to feed (SEE FIG-URE 12).

Also at each time step, a small fraction of the species on the lowest level go extinct. These individual extinctions, you might suppose, take place because of changes in the climate, the overzealous feeding of some predators, and so on. It doesn't really matter; biologists know that isolated extinctions can happen for all kinds of reasons. The important thing in the Amaral-Meyer game is what happens next. When a species on the lowest level goes extinct, this affects species on the level above. Some of these higher species may find that all the species on which they normally feed have vanished. In

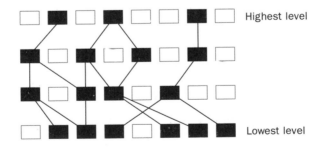

FIGURE 12. The Amaral-Meyer food chain. Each species at one level feeds on one or more from the level immediately below. Extinctions at the lowest level can trigger avalanches of further extinctions higher up the chain.

this case, the higher-level species will then also go extinct. This effect can ripple up the ladder, since higher species may also find that all their food sources have gone extinct.

In essence, the game fills out a food web randomly, allows a few species at the bottom to go extinct at any moment, and sends the effects of these extinctions rippling upward through the food web. The results of this trivial process are startling. The food chain naturally organizes itself into a critical state in which the extinction of just one inconspicuous species near the bottom can trigger an avalanche of further extinctions of any size. And most impressively, extinctions become precisely four times less likely each time the size is doubled, in excellent agreement with the real fossil record.

So it seems that the ordinary internal workings of ecosystems, in one form or another, may indeed lie behind the mass extinctions, and that critical-state organization is a core internal property of ecosystems. It seems to underlie not only their short-term ecological dynamics but their longer-term evolutionary workings as well, and gives rise to the peculiar patterns found in the fossil record. The apparent division of extinctions into mass extinctions and background events may very well be a fallacy.

There is, however, a niggling fly in the ointment. The author of an article in *New Scientist* once claimed that he had "yet to see any problem, however complicated, which, when you looked at it in the right way, did not become still more complicated."[8] He could have illustrated his point with the mass extinctions.

The External Reasserts Itself

To leave you with a fair picture of current thinking on the origins of the mass extinctions, I should mention the very intriguing results of yet one other theoretical game. In 1995, the physicist Mark Newman of Cornell University invented a simple game for extinctions in which the species do not interact with one another at all.[9] In other words, it dispenses entirely with internal ecosystem dynamics and supposes that the real cause of any extinction is an external shock to the ecosystem. As we saw in the last chapter, most paleontologists believe that such shocks in the form of climate changes, asteroid impacts, and so on are the true culprits behind the mass extinctions. Newman's model throws a very odd light on this old idea.

In the spirit of Bak and Sneppen, Newman tried to strip down the logic of extinction-by-external-shock to its most basic form. In his game, every species on Earth has a certain viability—a number between 0 and 1 that measures a species's ability to persist in the face of an external shock. Once again, we might picture a line of sticks. Newman imagines that a succession of shocks hit the Earth, and that each delivers a random stress to the species—numbers between 0 and 1. You can imagine each stress as a measuring stick held up to the collection of sticks. All those not tall enough get wiped out, and their positions are taken up by new sticks having a random length between 0 and 1. These are new species filling in the empty niche by ordinary evolution.

This continual culling of the ecosystem tends to wipe out most of the species with low viability—it makes the sticks longer. Newman supposes that in the periods between shocks, a small fraction of the high-viability species (corresponding to the tall sticks that were not culled and replaced) change their viability. The logic here is simple. Species do not evolve to be prepared for ice ages, asteroid impacts, or volcanic eruptions, but to be adapted to their immediate environmental conditions. There is no reason to think that becoming so adapted will also make a species more viable in the face of external upheaval. So, when evolution takes place between shocks, a species's viability will change randomly, getting a bit greater or smaller. Newman includes this effect by replacing a small fraction of the survivor sticks with new sticks of random length.

That's the complete game. You start with a collection of sticks of random length, and follow the rules. The only tunable feature of the game is the size of the shocks that hit the ecosystem. These are random, and yet you can choose the relative frequency of larger and smaller shocks any way you like. There could be many shocks that deliver stress close to 1, and only a few that deliver stress near 0, or vice versa. The remarkable thing about the game is that for almost any choice of the distribution of shocks, it works in the same way. The system organizes itself so that the result of the next stress applied to the system is unpredictable. Only a few species may go extinct, or almost all may do so. Indeed, the game gives rise to a power law for the distribution of extinction sizes that fits the fossil record quite well.

The profound importance of this simple game is that it shows how little the actual details of the distribution of shocks affects the record of extinctions. It is certainly true that changes in climate, volcanic eruptions, and occasional impacts of asteroids or comets do affect the world's biological communities. And yet very little is known about the exact consequences of this or that shock, or about how often and with what severity such shocks visit the Earth. We might despair of ever knowing their long-term consequences for life on Earth. But in the light of Newman's work, it seems we can estimate their long-term consequences even without knowing much about the shocks that actually hit the Earth. We get something for nothing.

So when it comes to the mass extinctions, everything leads to spreading confusion. In view of Newman's game, there is no way to know for sure whether the mass extinctions were triggered by external or internal shocks. One of the aims of current research in this area is to explore these games further, to see if subtle differences in their mathematical signatures can be detected in the fossil record, and so settle the matter. But even if Newman's game prevents us from making a bold final conclusion about the mass extinctions, it serves to emphasize the precarious state in which the global ecosystem appears to live—always on the edge of radical upheaval.

Terrible Science?

But it is high time we face up to an objection that might be leveled against all the work we have touched on in the preceding several chapters. Stretching back as far as the freezing game of chapter 3, we have been playing fast and loose with the details of this or that problem, glibly throwing them out, and hoping that the skeletal games that result really still might offer a legitimate explanation of how things work. Isn't this a rather reckless way to do science? Are there any good reasons to expect that simple games can capture the essence of freezing, let alone of earthquakes, forest fires, and mass extinctions?

It is fair to point out that Bak and Tang's initial paper describing their earthquake game touched off a storm of criticism. Many geophysicists have spent their careers studying specific earthquake zones and fault systems in painstaking detail in an effort to understand earthquakes. To them, this slapdash mathematical approach seemed almost insulting, and served only to confirm about theoretical physicists what the biologist Francis Crick once said about mathematicians. "In my experience," he concluded, "most mathematicians are intellectually lazy and especially dislike reading experimental papers."[10] After all, here were a handful of theorists who never took a first university course in Earth science, and who were nonetheless claiming that they could explain earthquakes with a toy model that includes almost nothing of the complex detail of the physical setting in which real earthquakes occur.

The model of Burridge and Knopoff is already an outrageous oversimplification of the real problem, as it throws out almost all details about the geometry and physical properties of real faults. Earthquakes happen in real rocks. Yet the game doesn't even mention the properties of rocks, except to concede that they are elastic, as reflected, almost as in afterthought, in the springs used in the game. Moreover, earthquakes in the real world only rarely involve a single fault. They almost always involve highly complex networks of faults, and so you cannot say that a quake happened on this or that fault. Yet the earthquake game includes only one fault. Bak and Tang's pared-down version is even worse, as it willfully violates the laws of physics—the inevitable effects of friction—in following the movement of the blocks. The Olami, Feder, and Christensen version partially rectifies this problem, but it is hardly based on a close reading of the physics. How can intentional falsification of what is known to be true physics possibly lead to any valuable insight into real earthquakes? All these models might be cute games, these geophysicists concluded, but they are merely games. And the games' curious agreement with the Gutenberg-Richter law can only be a meaningless and uninteresting coincidence.

One might point out that Isaac Newton would never have understood the Earth's motion around the Sun had he not ignored every last detail about the Earth except one: that its mass is affected by gravity. He didn't worry that it has a core and a mantle, or that each day, because of the tides, oceans of water slosh back and forth on the surface. He didn't worry that, strictly speaking, the exact position and mass of every last tree on the planet should come into his calculations. Newton assumed that none of this would make much difference, and he was right.[11] Even all that sloshing water doesn't alter the length of the year by a single minute.

Nevertheless, this objection over the omission of detail needs addressing. The forest fire game doesn't even attempt to describe the mosaic of different kinds of trees that make up a forest, but instead treats all trees as if they were identical. They all catch fire just as easily, and burn at the same rate. Nor does the game include any of a forest's natural barriers to fire, such as rivers or roads. The game ignores firefighters and the influences of weather. Indeed, one could easily fill a book with the details of real forests of which this game makes no mention.

In view if this situation, many scientists continue to object to the very approach represented by all these games. In their eyes, this is simply a terrible way to do science. Does this objection hold merit? To see, we might try to reach ever more deeply into the details in each of these settings, hoping to tease out those that really matter from those that do not. But the wealth of detail is nearly infinite, and, fortunately, there is another way to answer the objections about the willful neglect of reality.

There is a powerful scientific understanding of the critical state that goes far back before the sandpile game and its other simple cousins. Indeed, there are deep theoretical arguments showing that it is often valid to be reckless with the details, and that the workings of outrageously oversimplified games really can offer legitimate explanations of very complicated things. The basic idea goes by the name *critical-state universality*, and it represents one of the most profound discoveries in theoretical physics in the twentieth century.

De Magnete

We live today in a world in which poets and historians
and men of affairs are proud that they wouldn't even begin
to consider thinking about learning anything of science,
regarding it as the far end of a tunnel too long for
any wise man to put his head into.

—J. ROBERT OPPENHEIMER[1]

Basic research is like shooting an arrow into the air and,
where it lands, painting a target.

—HOMER ADKINS[2]

THE RUSSIAN PHYSICIST PYOTR KAPITSA, DIRECTOR FOR THREE decades of the great Institute of Physical Problems in Moscow, was once asked during a trip to England about the meaning of a crocodile that someone had painted on the side of a laboratory of the Royal Society. He stood back and studied the painting and, at length, concluded that it should be properly interpreted as a statement about the nature of science. "The crocodile cannot turn its head," he noted. "Like science, it must always go forward with all-devouring jaws."[3]

In his laboratory in Moscow in 1938, Kapitsa cooled helium gas down to the incredible temperature of minus 271 degrees Celsius, just two degrees above absolute zero, the coldest of all possible temperatures. He discovered that as helium gas becomes ultracold, it turns first into an ordinary liquid, and then, at minus 271 degrees, into one of the strangest substances ever encountered: a superfluid. A superfluid can be poured or held in a jar like any liquid, but if you set it swirling in a bowl, it will swirl forever.[4]

This sudden change in the state of helium is an example of what physicists refer to as a *phase transition*. When ice melts in a gin and tonic, or when a puddle evaporates into the air, these too are phase transitions: each being the transformation of a substance from one form, or "phase," to another. In every phase transition, there is a change in the internal workings of the stuff, as its atoms or molecules organize themselves differently. An iron magnet at room temperature can pick up nails and tug on horseshoes, but as an English physician and scientist named William Gilbert discovered in 1600, a magnet heated in a furnace eventually loses its magnetic power—yet another phase transition.

One would think that the all-devouring jaws of science would many years ago have swallowed up any mysteries about iron magnets—and about phase transitions more generally. But in truth it took physicists nearly four centuries to fathom what happens in a substance as it goes from one phase to another, and especially to comprehend that delicate state of matter that results when a substance is held on the knife-edge between one phase and another. This is where scientists first encountered the critical state: Kapitsa's helium balanced precariously at minus 271 degrees, where it is both ordinary liquid and superfluid at once; an iron magnet poised on the thin boundary between its magnetic and nonmagnetic conditions.

But several centuries' work was well worth it. For in the late 1960s physicists discovered that there is a universal organization for substances in a critical state. Remarkably, while there are thousands of different substances all complicated by their own specific details, the critical state is in every case much the same. So the world is immeasurably simpler than it might have been. The lesson is this: When it comes to understanding something in a critical state, most of the details simply *do not matter.* Physicists refer to this considerable miracle as *critical state universality,* and the principle has now been supported by thousands of experiments and computer simulations. In this profound principle, we shall find a powerful answer to those objections of the last chapter. Let's take a closer look at what goes on inside a magnet.

Coming to Order

Every atom in a chunk of iron is itself a tiny magnet, and can point in any direction: up, down, left, right, etc. You might imagine the inside of a piece of iron as an army of arrows. Physicists knew even a century ago that whether a piece of iron is magnetic or not has something to do with the organization of this army. The iron might be sitting on a table at room temperature, or be piping hot in a furnace. The crucial question is this: Where do all the arrows point?

Being what they are, the atomic magnets would like to line up with one another. If left to themselves, they would do so, falling quickly into formation like any well-disciplined army. But the arrows have a disrupting enemy to contend with: heat. The temperature of anything measures how much disorganized energy there is in it; in warm air, the molecules fly about more violently then they do in cold. In solid iron, the atoms do not fly about, but quiver about fixed positions, the vibrations becoming ever more violent as the iron gets hotter. So while the magnetic forces between the iron atoms try to line them up, heat treats them to a storm of disrupting abuse. There is a war between the forces of order and of chaos, and its outcome determines how a magnet behaves outwardly.

If the iron sits on a table at room temperature, the jostling of the atomic magnets is fairly weak, and they succeed in lining up. The strength of each atomic magnet is extremely tiny, of course, but the number in even a small chunk of iron is well over 10^{24} (that is, 1,000,000,000,000,000,000,000,000). Working together, this army amounts to something, and the iron can pick up nails. If the iron sits glowing in a furnace, on the other hand, then an annihilating storm of noise will overwhelm the forces of order. A snapshot now would show an army in disarray. In this case, the effects of all the tiny magnets cancel out, and the iron cannot pick up nails (SEE FIGURE 13).

The story, then, seems rather simple: when iron is cold, the forces of order win out; when hot, the battle goes the other way and chaos rules. But this leaves out the juiciest detail. At some intermediate temperature, the forces of order and chaos must fight to a stalemate. This is the critical point, and in iron, it occurs at 770 degrees Celsius. What happens to the army of arrows at this point? What does it mean for something to be neither organized nor disorganized, but somehow perched on the delicate boundary between the two? The answers to these questions are rather more elusive.

The theory that physicists use to understand gases and liquids and other things made out of huge numbers of atoms or molecules is known as statistical mechanics. But using this theory to build a

(a)

(b)

FIGURE 13. At high temperatures (*a*), the atomic magnets within a piece of iron cannot line themselves up. The abusive storm of thermal jostling is simply too much. But if the temperature falls below a certain critical level (*b*), the battle goes the other way: now facing a weaker storm, the magnets manage to organize and the iron becomes magnetic.

detailed theory of a real magnet is all but impossible. Even when confronted with the army of atomic magnets as we've been drawing them, the mathematical machinery of the theory quickly grinds to a shuddering halt. As a result, the first physicist even to catch a glimpse of what goes on at the critical point was the Norwegian Lars Onsager, in the 1940s. And the only way Onsager could make progress was by simplifying the picture of a real magnet to the point of absurdity.

To picture Onsager's magnet, imagine that the individual arrows have been yoked so that they can point only up or down. Suppose also that each magnet, rather than influencing all others, can affect only a handful of those sitting next to it. Finally, suppose we imprison this magnet in "flatland," in just two dimensions, forcing it to live like an insect pressed between two sheets of glass. You might picture an array of atomic magnets laid out on the squares of a chessboard. This is, of course, a ridiculously oversimplified picture of any real magnet. But the simplifications allowed Onsager, with the aid of a gifted student named Bruria Kaufman, to gain a clue to the nature of the organization that lives within a magnet when the magnet is tuned to the critical point.

Suppose some magnet X in the midst of the crowd just happens to be pointing up. How does this affect another magnet some distance away? In physics vernacular, this is a question about the "correlations" between magnets, and for the toy magnet poised at the critical point, Onsager and Kaufman found a curious result. If the pattern of magnets were completely random, then any two would have exactly half a chance of pointing in the same direction (remember, they can only point up or down). The exact calculation showed instead that the closer two magnets are, the more likely they are to line up with each other.

This seems reasonable, and not too profound. Every time Onsager and Kaufman doubled the distance between a pair of magnets, the slight bias those magnets had for pointing in the same direction decreased by a factor of about 1.19. This pattern held for magnets separated by 10 or 100 or 1,000 squares on the chessboard, and held just as well for those separated by 100,000 or 100,000,000. In other words, the result was a power law. What does it mean?

The Rise of Factions

It is for good reason that the digital computer has been called "the most important epistemological advance in scientific method since the invention of accurate timekeeping devices."[5] Onsager and

Kaufman couldn't use it, of course, but we can, and it offers by far the easiest way to cut to the core meaning of their power law. With a fast computer it is not hard to set up something like 250,000 atomic magnets on a grid, and let things rip. The idea is to run the calculation for a variety of temperatures, and to display the resulting patterns by coloring the "up" magnets white and the "down" magnets black. Let's look at some typical results (SEE FIGURE 14).[6]

The first two images (FIGURE 14, A AND B) correspond to the magnet above or below the critical temperature. As expected, thermal noise wins out above the critical point, and here the magnets lie in disarray. Each flips back and forth rapidly and at random; this is the realm of pure chaos, and the image looks like the static on a badly

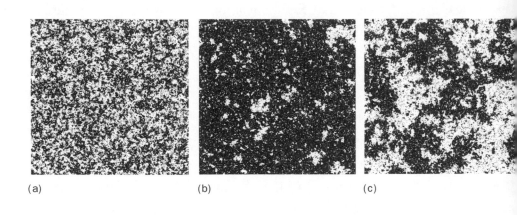

(a)   (b)   (c)

FIGURE 14. Above the critical point (*a*), "flatland" is in chaos. The atomic magnets point up (white) and down (black) with equal likelihood, and where one magnet points says nothing about where its neighbors may be pointing. Below the critical point (*b*) lies the regime of order, and here almost all the magnets line up with one another. The critical point itself (*c*) is a peculiar netherworld between order and chaos, where white and black magnets mingle in ever shifting factions of all sizes. Images from J. J. Binney et al., *An Introduction to the Theory of Critical Phenomena* (Oxford University Press, 1992), reprinted by permission.

mistuned television. At a temperature below the critical point, on the other hand, almost all of the magnets have succeeded in lining up, in this case in the "down" direction, and in this regime of order, we see a field of nearly complete darkness.[7] Nothing surprising here.

Now take the second image and increase the temperature just a bit, moving toward the critical point. Things begin to get more interesting: while most of the magnets stay aligned as before, a few white pockets of dissenters begin to invade. Increase the temperature still further, and these pockets grow and become more numerous. Eventually, as we approach the critical temperature (FIGURE 14, C), the factions of dissenting magnets grow to such an extent that one could walk across the entire array without ever leaving the white or, if you like, without ever leaving the black. This is the critical state: the condition of the magnet balanced precisely between its magnetic and nonmagnetic states. And the image reveals the meaning of the power law.

At this critical point, the factions of magnets come in all sizes, ranging from single isolated magnets up to huge clumps that stretch across the entire magnet. If we added more magnets to make an array as big as the United States, the same thing would happen. The factions would range from tiny clumps to massive blobs stretching from New York to Los Angeles. As we know, the geometric regularity of any power law implies a lack of any typical scale—a feature that shows up clearly in the critical point image. But one picture really can't do justice to the character of the critical state, which is forever changing.

If you were to take snapshots at different moments, you would see the alliances of the factions constantly shifting, with some dissolving and others forming up. The critical state is subject to tremendous fluctuations, and always remains poised on the very edge of sudden, radical change. To call it "hypersensitive" would be an understatement. Since the army of arrows is balanced on the precipice between its two phases, always on the verge of falling into line, even the tiniest influence can push it over the edge. Just a single magnet flipping over can trigger an avalanche of further flippings that would rampage from one side to the other.

But wait a minute, let's not get carried away. The computer images we've been looking at come from simulations of the same silly toy—a magnet living under torture in "flatland." If we want to understand real iron magnets, we can hardly expect this gross parody of reality to give us trustworthy clues. What happens if we add a good dose of realism back into the picture? What about real iron magnets? And what about other substances, such as Kapitsa's helium? To answer these questions, we need to look at a few critical numbers.

Deep Principles

The number 1.19 that appears in the pattern of Onsager and Kaufman's power law is a kind of mathematical label for this particular critical state. In the power-law patterns of earlier chapters, we found in each case an equally regular pattern when moving from one scale to another, and yet the precise numbers changed from one case to the next. Double the size of an earthquake, and the Gutenberg-Richter power law says those quakes become four times less frequent. Double the size of the branch you are looking for in a cluster of the freezing game, and the power law says those branches will be about three times more numerous. Each such number corresponds to one particular kind of self-similar pattern. To be more specific about the character of the critical state, physicists pay attention not only to the form of the power law, but to the numbers that appear in it.

Since there can be many different critical states, each with a different value of its critical number,[8] we might expect these different critical states to show up in phase transitions involving different substances, whether these be real or, like Onsager's magnet, imaginary.

Since Onsager's silly toy has the number 1.19, we should expect the real iron magnet to have another number. After all, Onsager ignored reality almost completely in making up his model. In general, we should expect a different critical number for each phase transition. The interactions between the atoms in a gas or liquid are not even remotely like those between two tiny magnets. Atoms or molecules zoom about and collide with one another, while magnets stay put and simply change their orientation. In the case of superfluid helium, the rules of the quantum world play an important part, and the forces at work are not even forces as we ordinarily know them.

Thus researchers in the 1960s studying the changes from vapor to liquid of substances as diverse as oxygen, neon, and carbon monoxide were mystified when they found exactly the same critical numbers. They were even more surprised when these very same numbers turned up also in phase transitions involving the mixing and separation of chemicals, a situation bearing no similarity whatever to the change of a vapor to liquid. Most shockingly, the same numbers showed up yet again in calculations for the three-dimensional version of the toy magnet that Onsager had studied—an extremely crude model for a magnet, and most obviously one having nothing even remotely to do with mixing chemicals or condensing liquids.

By 1965, physicists were staring into the face of an almost unbelievable possibility: not only does the critical state and the rise of factions appear in every phase transition, but the precise mathematical character of this state depends in almost no way on the details of the things involved.[9] This idea remained only a tantalizing and half-formed possibility, until in 1970 a young physicist from the University of Chicago put it on more solid ground. If few details seemed to matter, Leo Kadanoff managed to put his finger on those few that really do.

At the critical point, pockets of organization are just about to break out at any place at any moment, and are continually breaking out, as factions grow and then disappear. How large do the factions grow? How quickly do they dissolve? These questions are all down to the basic geometrical issue of how easy it is for an ordering influence at one point to bring similar order to another nearby. This is not physics, but geometry. In an ordinary magnet in three dimensions, any atomic magnet can reach out to influence its neighbours in three independent directions. In "flatland," on the other hand, one of these possible directions has been pressed out of existence.

In studying the critical numbers for different phase transitions, Kadanoff found that the basic physical dimension of the thing in question—of the very space in which it lives—is one of the factors that matter. He also found that only one other detail seems to matter, this being what you might call the "shape" of the individual elements. In a gas of xenon, for example, each atom is like a tiny billiard ball. It can move around, but it can't point. In a magnet, the atoms are like arrows, and can "do" more since they can potentially point in lots of directions. When the individual elements have more options, you can imagine that it is harder for order to propagate from one place to another. Sure enough, this detail also affects the precise form of the self-similarity in the critical state.

Incredibly, however, Kadanoff found that nothing else seemed to matter.[10] So forget the atomic masses and the electrical charges of the particles involved. Forget whether those particles are atoms of oxygen, nitrogen, krypton, nickel, or iron. Forget even whether they are made of single atoms or are more complicated molecules made of several or even a hundred atoms. Forget everything, in fact, about the kinds of particles and how strongly or weakly they interact with one another. None of these details affects the organization of the critical state even a tiny bit. This is critical state universality.

In the critical state, the forces of order and chaos battle to an uneasy balance, neither ever fully winning or losing. And the character of the battle, and the perpetually shifting and changing strife to which it leads, is the same regardless of almost every last detail of the things involved. The physical dimension of the thing in question matters, as does the basic shape of its elements—points, arrows, and so on. But nothing else matters.

Let's take a tiny but informative step into the abstract, and imagine the conceptual world of all conceivable substances. This world will divide up naturally into nations. There will be the nation of "arrowlike things living in three dimensions," and another of "pointlike things living in just one dimension," and so on. Physicists refer to these nations as *universality classes*.[11] The miracle of universality is that any two substances, real or imaginary, that fall into the same class will necessarily have exactly the same critical-state organization, regardless of how utterly dissimilar they may otherwise seem to be.

Critical Thinking

We have taken this brief detour in order to meet the critical state in more detail, and to understand its origins and its peculiar properties. And we are now ready to confront its deepest and most profound implication. For in the principle of critical-state universality, nature has given scientists an amazing gift.

Since all physical systems fall into universality classes, if you succeed in understanding the critical state in any system from one class, you have immediately understood all systems in that class. But note that the crudest sorts of toy models also fall into these very same universality classes. So to understand any real physical system in a critical state, you may as well forget all the real, messy details about that system and focus instead on the simplest mathematical game belonging to the same universality class. It can be crude, even ridiculously so. It can break the laws of physics and ignore virtually every detail of the real system, and yet you have a guarantee that it will have exactly the same critical behavior as the real physical thing, so long as those two crucial dimensions are correct. Even hideously crude models can work exactly like the real thing.[12]

This brings us back to all those objections we faced at the end of last chapter. The earthquake games we looked at in chapter 3 bear hardly any relation to the earth's real crust. No property of even a single real rock enters the model, nor does the model respect the real-world truth that earthquakes happen on networks of faults, rather than on single faults. The forest fire and evolution games also offer absurdly crude models for reality. How can intentional falsification of what are known to be true facts possibly lead to any valuable insight? If these toy models give power laws much like those observed in reality, is this anything more than meaningless coincidence?

Well, for something in a critical state, it is possible to understand its essential organization while neglecting almost every single detail, as long as we do not neglect a few really crucial details. This undermines objections over lack of detail. It is indeed possible to capture the essential workings of the earth's crust, forests, and ecosystems in terms of incredibly crude models.[13]

Thus we have arrived at what we might call an attitude of critical thinking. Things that live in critical states tend to show similar kinds of organization, and this organization arises not from specific details of those systems and the elements that make them up, but from the far deeper skeleton of basic geometry and logic behind these details. The critical form wells up in things regardless of what they are. When something is recognized to be in a critical state, its essential character can be understood even by ignoring most of the details.

With the conceptual tools we have gathered so far, we can now begin asking what all this implies about the nature of the human world. Unfortunately, it is not often easy to put precise numbers on social changes. Political revolutions and waves of new fashion affect us all, but they cannot easily be measured with the same precision as fluctuations in a magnet, or vibrations in the Earth's crust. So we might well begin by looking to the financial markets, where the prices of stocks and bonds have been recorded every few seconds for decades, and where there is indeed a staggering wealth of data.

·8·

Wild at Heart

Year after year, economic theorists continue to
produce scores of mathematical models and to explore in
great detail their formal properties; and the
econometricians fit algebraic functions of all possible
shapes to essentially the same sets of data without
in any way being able to advance, in any perceptible way,
a systematic understanding of the structure
and the operation of a real economic system.

—WASSILY LEONTIEF[1]

Zauberman's Law: The worse the economy,
the better the economists.

—ALFRED ZAUBERMAN[2]

✦

"AN ECONOMIST," ACCORDING TO A JOKE THAT ECONOMISTS TELL among themselves, "is a trained professional paid to guess wrongly about the economy." Another joke claims that economists have predicted "all nine of the last five recessions." Neither would be funny, of course, if it didn't contain an element of truth. In 1995, the independent economics consultancy London Economics compared the recent prediction scorecards of more than thirty of the top British economic forecasting groups, including the Treasury, the National Institute, and the London Business School. John Kay of the London Business School summed up the consultancy's findings for the *Financial Times:*

> It is a conventional joke that there are as many different opinions about the future of the economy as there are economists. The truth is quite the opposite. Economic forecasters . . . all say more or less the same thing at the same time; the degree of agreement is astounding. The differences between forecasts are trivial relative to the differences between all forecasts and what happens. . . . what they say is almost always wrong. . . . the consensus forecast failed to predict any of the most important developments in the economy over the past seven years—the strength and resilience of the 1980s consumer spending boom, the depth and persistence of the 1990s recession, or the dramatic and continuing decline in inflation since 1991.[3]

There is, of course, nothing special about the British economy that might make it particularly opaque to prediction. Nor are the economists at these institutions inept. All over the world, year after year, economic prognosticators of all nationalities make their contributions to what can only be called a concert of prediction failure. In 1993, the Organization for Economic Cooperation and Development (OECD) analyzed forecasts made between 1987 and 1992 by the governments of the United States, Japan, Germany, France, Italy, and Canada, as well as by the International Monetary Fund (IMF) and the OECD itself. The conclusion? Not only were each of these organizations' predictions abysmally inaccurate, but they would have made better predictions for inflation and gross domestic product if they had scrapped all their "sophisticated" economic models and simply guessed that the numbers in each year would be unchanged from the last.[4] Surveying predictions over the past century, a respected financial analyst recently concluded: "When they are predicting anything that involves money, economists, prominent investors and the reporters who quote them haven't been wrong on occasion: they've been unerringly errant."[5]

All the same, mainstream economists seem to share a faith in the ability of governments and central banks to steer the economy by adjusting "policy levers." In the pages of *The Wall Street Journal* or *The Financial Times*, economists, businessmen, and government finance ministers argue endlessly about how best to tweak public spending, tax rates, and so on. There may, of course, be some truth to this idea: If the Federal Reserve Bank were to raise interest rates by a few percentage points tomorrow, it seems a good bet that this would act as a brake on the U.S. economy. Similarly, lowering taxes should boost consumer spending.[6] On the other hand, it is easy to find economists who sometimes take their optimism about economic controllability a bit too far. In 1998, an economist from the Massachusetts Institute of Technology, writing in reference to the buoyant economy of the United States, claimed that

> this expansion will run forever; the US economy will not
> see a recession for years to come. We don't want one, we

don't need one, and therefore we won't have one. . . . we
have the tools to keep the current expansion going.[7]

It may be bad manners to note that, as of the early months of 2001,
the U.S. economy appears to be slipping into recession. More generally, however, if steering the economy is easy, why have we suffered
through recessions in the past? And why aren't economists better at
predicting its movements? How could an event as tumultuous as the
1987 stock market crash arrive without any warning? Indeed, how
could it happen at all?

Twinned with the prevailing conviction that economies can be
steered is an equally strong belief that any sudden and dramatic
swing one way or another must have some specific, identifiable
cause. In the case of the 1987 crash, as we saw in chapter 1, many
analysts pointed to computerized trading, just as others pointed to
excessive borrowing as the cause of the great 1929 crash. In 1997,
and again in an explanation offered after the fact, economists pointed
to massive foreign debt as the cause of the sudden collapse of an economic miracle: the incredible (until then) Tiger economies of
Southeast Asia.

Given the sheer number of people taking part in any economy,
all with their own individual ideas and strategies, hopes and apprehensions, and the similarly enormous number of companies and
organizations, all competing with their own separate aims, perhaps it
is no surprise that predicting the economic future is difficult. Then
again, we have seen similar patterns in preceding chapters. So before
we throw up our hands and attribute all this intransigence to the
human factor, it makes sense first to see if there might not be a simpler explanation. As we have seen, great earthquakes, forest fires, and
mass extinctions are all merely the expected large fluctuations that
arise universally in nonequilibrium systems. To avoid them, one
would have to alter the laws of nature.

Admittedly, to venture into economics is to leave the laws of physics and biology far behind. You cannot capture with mathematics the intelligence or the whims of a real person, let alone millions of people, or even mimic in the most rudimentary way their emotions, dreams, and desires. Unlike magnets or fragments of the Earth's crust that move according to strict physical rules, people make choices. Still, ideas, feelings, desires, and expectations can be infectious, and just as one microscopic magnet can affect the alignment of a neighbor, so can the actions of one person or business firm influence another. And what do we know about the critical state? We know that its organization depends in almost no way on the precise nature of the things involved, but only on the way that influences can propagate from one thing to the next.

Might the critical state be lurking within the workings of human social systems as well? Strings of numbers come pouring from Wall Street every day, and since the signature of the critical state is a mathematical pattern, the financial markets offer a good place to start looking for it.

Down to Fundamentals

Behind the widespread belief that the economy can be sensibly tweaked, adjusted, and controlled lies a core idea of economic theory known as the *efficient market hypothesis*. If one economist recently called it "the most remarkable error in the history of economic theory,"[8] many others continue to accept it with seemingly little hesitation.[9] The idea proclaims that everyone in the market always acts in his or her own greedy self-interest. What's more, as the authors of a recent paper describe it,

> unable to curtail their greed, an army of investors aggressively pounce on even the smallest informational advantages at their disposal, and in doing so, they incorporate

their information into market prices and quickly eliminate the profit opportunities that gave rise to their
actions.[10]

If a stock is undervalued, the idea goes, people will quickly buy it up
since they can make money by selling it later. As demand rises, the
price does also, until the stock is no longer cheap and equilibrium
has been restored.

In an efficient market, supply matches demand perfectly, and
prices always have their proper values—that is, values matched to the
underlying "fundamentals." If you own stock shares, you get paid
dividends, and the stock's authentic value—how much a reasonable
person should be willing to pay for it—should depend on the company's realistic prospects for growing, making profits, and paying out
fat dividends in the future. So stock prices on the New York Stock
Exchange should reflect the fundamental facts about the real values
of stocks. If a company commits a blunder, or if laws change that
put it at a competitive disadvantage, then the fundamentals have
changed, and its price should fall.

Economists readily admit that on the basis of this view, market
prices should bounce up and down gently and erratically. No one can
predict how prices will change, as it is information hot off the press
that triggers such changes by influencing the underlying fundamentals. What new technological advance or wrinkle in corporate policy
lies just around the corner? Nevertheless, as new information trick-
les in, prices should change accordingly and always bring the economy back into equilibrium—that is, into balance.

In this conventional view, the economy is something like a bath
of water. On the microscopic level, what's going on in is a terrible
mess, with individual molecules doing all sorts of crazy things. But in
equilibrium, all this microscopic nonsense gets averaged out. Tilt the
bath and it is easy to predict how the water will rearrange itself as it
seeks equilibrium in accordance with the laws of physics. In eco-

nomics, similarly, pulling a lever to decrease interest rates, for example, should tilt the playing field for every rational person in the market. With money less costly to borrow, each person or company should borrow and spend a bit more. This should stimulate the economy, which should quickly settle into a new equilibrium with, for instance, higher output.

But there is one problem: no amount of equilibrium thinking can account for such huge and rapid fluctuations as the stock market crashes of 1929 and 1987. What made the Dow Jones Industrial Average lose more than 22 percent of its value in one day in 1987? The Dow is an average of a handful of stocks for companies spread over a range of industries and chosen as good indicators of general economic vitality. As one economist points out,

> . . . it is difficult to believe that there could be a sudden change in the fundamentals which would lead agents simultaneously within half a day to the view that returns in the future had gone down by over 20 percent. Yet this is what would have to be argued for the October 1987 episode on the world stock markets.[11]

Faced with this implausible idea, most analysts blame the computer trading mentioned earlier. Many seem convinced by this ad hoc explanation, and even feel certain that another crash cannot happen because the problem has been fixed. In 1998, referring to such computerized trading programs, two prominent economists wrote in *The Wall Street Journal* that "sources of structural fragility have been substantially, if not totally, corrected. . . . a repeat of the hair-raising events of 1987 [is] highly unlikely."[12]

But mathematical research over the past decade tells a very different and less comforting story. According to the numbers, sudden upheavals are very far from being highly unlikely, and may indeed even be inevitable. In direct conflict with everything the efficient market hypothesis stands for, large fluctuations in market prices seem to result from the natural, internal workings of markets, and so flare up from time to time even if there aren't any "sources of structural fragility," or sudden alterations of the fundamentals to set them off. And the reason may be quite simple: markets are not even remotely close to being in equilibrium.

Furious Fluctuations

In 1900, a Frenchman named Louis Bachelier presented to the faculty of the École Normale Supérieure in Paris a curious dissertation entitled *Théorie de la Spéculation*. Bachelier's dissertation was not well received by his faculty, and he was eventually blackballed in his attempts to find an academic position.[13] Perhaps his professors were simply disappointed that he had not focused on one of the traditional topics of theoretical or experimental physics. Instead, Bachelier had attempted to formulate a mathematical theory for price movements.

Suppose that today the price for a pound of cotton is $10. What will it be, say, one month later? There is no way to know for sure; this is a matter for statistics and probabilities. Bachelier supposed that if you recorded the changes in cotton prices over many one-month intervals, those changes would fall onto something like the bell curve that we met in chapter 3 (SEE FIGURE 1). Since the bell curve fits so many things in nature, this seems like a good guess. Overall, the price would go up as often as down, and so the hump of the curve should straddle zero, "no change" being most likely. Recall that the tails of the bell curve fall off very quickly. This fit in with Bachelier's belief that price changes greater than some typical size ought to be extremely rare.

Overall, then, Bachelier viewed prices as following a gentle "random walk," the numbers each month hopping up or down by a typically small amount, and this mathematical picture produced charts with a startling resemblance to charts of real prices. Bachelier certainly deserved a more respectful treatment from his faculty committee. Indeed, the first signs of trouble with his theory of price fluctuations didn't crop up for more than half a century.

In 1963, IBM mathematician Benoit Mandelbrot was perusing the record ups-and-downs in cotton prices when he stumbled over a peculiar pattern: self-similarity. Any small segment of a record of price changes, if stretched out, ended up looking again much like the whole. This is one way to see self-similarity, but Mandelbrot also found another. Bachelier was right in his belief that the changes in prices up or down are indeed random. Mandelbrot found that if the price goes up a bit this month, that doesn't mean it is any more or less likely to go up again, or to fall, next month. Prices really do follow a random walk. But when Mandelbrot looked at how those random changes were distributed by size, he found nothing at all like Bachelier's bell curve.[14] Instead, he discovered that price changes follow the form of a power law. So despite what Bachelier assumed, price changes don't have a "typical size."

In the past decade, researchers have used computers to take a far more exhaustive look at fluctuations in stock and foreign exchange markets all over the world, in every case finding a similar story: power laws and wild fluctuations having no inherent scale. In 1998, for instance, Gene Stanley of Boston University led a team of researchers in analyzing the fluctuations in the famous Standard & Poor's 500 stock index.[15] Based on the share prices for five hundred large corporations on the New York Stock Exchange, this index is a kind of single benchmark for the entire market. Stanley and his collaborators studied prices recorded every fifteen seconds over thirteen years from 1984 to 1996—an incredible 4.5 million data points in all. The index over these years shows a long, slow, increasing trend complicated by lots of irregular ups and downs, as in figure 15a below.

To bring out the fluctuations, we can simply ignore the trend, and also ignore whether prices went up or down over any interval. This leads to a more illuminating graph showing only the total size of the price change that took place over every one-minute interval, shown in figure 15b. The image clearly has a very "spiky" character. Looking at these price fluctuations in detail, Stanley and his colleagues found that price changes become about sixteen times less likely each time you double the size. Recall that it is not so much the number in this power law but the regular geometric form that is so important, for it means that there is no qualitative difference between the larger and smaller fluctuations.

This power law implies that there is no such thing as a typical fluctuation, and so no reason to think that the largest swings either up or down are in any sense unusual. The idea that sudden and terrifically large changes need explaining doesn't seem to hold up; in conflict with our intuition, even these are merely business as usual. Scientists sometimes refer to power-law distributions as having "fat tails," because in comparison to the ends of a bell curve, the tails of a power-law curve don't fall off so quickly. The tails of the distribution correspond to extreme events, and when something works to the tune of a power law, extreme events are not really so unlikely. Indeed, it is even a misstatement to refer to them as "extreme."

A similar power-law form holds for fluctuations in the Standard & Poor's index over minutes, hours, and days, and also in the prices of the stocks of some one thousand individual companies.[16] Other researchers have found similar power-law fluctuations in other stock markets[17] and foreign exchange markets,[18] so wild fluctuations seem to be a universal feature of all kinds of markets. And the study of "raw" price changes is only one way to bring this out.

Stanley's team has also turned its attention to market "volatility,"

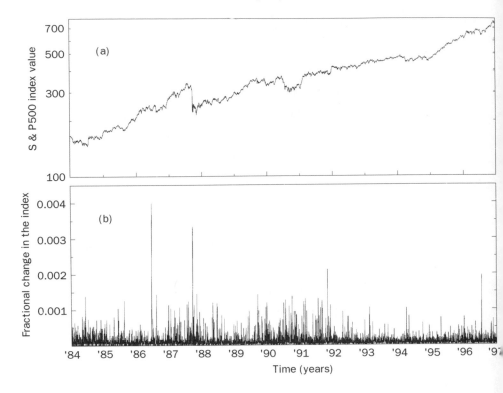

FIGURE 15. The minute-by-minute values (*a*) for Standard & Poor's 500 stock market index (S&P 500) between 1984 and 1997 show numerous fluctuations on top of a gradual upward trend. To bring the fluctuations into relief, it helps to ignore the trend and disregard whether prices went up or down over any interval. This leads to a graph (*b*) showing only the total size of the price change (in percentage terms) that took place over every one-minute interval. Images adapted from Y. Liu et al., "Statistical Properties of the Volatility of Price Fluctuations," *Physical Review E* 60 (1999): 1–11, reprinted by permission.

a measure of the wildness of prices and another quantity of great interest to stock market investors. The idea is to take the record of minute-by-minute price changes and break it up into "windows" of several hours. You can then work out how wild and vigorous the minute-by-minute price fluctuations were within each window, and so get an idea of how the fluctuations themselves grow stronger or weaker. A plot of these numbers for the Standard & Poor's index shows that the market is far calmer at some times than at others (SEE FIGURE 16). To take matters one step further, you can even look at the fluctuations in this volatility—that is, at how vigorously and erratically the market flits between its times of wildness and calm. Here again, researchers have found a scale-invariant power law: the market doesn't have a typical wildness in its fluctuations. Even its volatility is itself highly volatile.[19]

All these wild fluctuations fly in the face of the efficient market hypothesis, and the notion that markets are in equilibrium. In equilibrium, the fluctuations should be like Bachelier thought they were—small. What could make them so big?

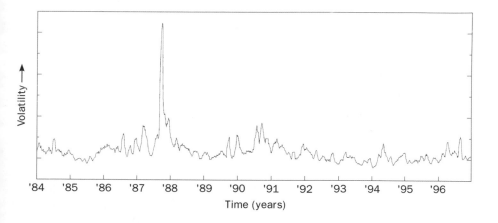

FIGURE 16. Averaging the sizes of the minute-by-minute price changes (from figure 15b) over month-long intervals offers a way to see if the market fluctuates more violently at some times than at others. It most certainly does. Even the vigor of the fluctuations varies wildly. Image adapted from Y. Liu et al., "Statistical Properties of the Volatility of Price Fluctuations," *Physical Review E* 60 (1999): 1–11, reprinted by permission.

The Crowd

To anyone who is not an economist, the orthodox perspective that sees people as "rational agents" who always work out their rational self-interests and act on them seems more than a little peculiar. As described by the economist Paul Ormerod,

> In orthodox economic theory, the agents involved in any particular market . . . are presumed to be able to both gather and process substantial amounts of information efficiently in order to form expectations on the likely costs and benefits associated with different courses of action, and to respond to incentives and disincentives in an appropriate manner. . . . The one thing these hypothetical individuals do not do . . . is to allow their behaviour to be influenced directly by the behaviour of others . . . and their tastes and preferences are assumed to be fixed, regardless of how others behave.[20]

To bring home just how ill conceived this view is, consider the multibillion-dollar advertising industry. It would fly well past the limits of naïveté to suggest that advertisers simply offer us information so that we can make better decisions. Advertising works because we can all be influenced and manipulated, and, what's more, because, once we have been influenced, our beliefs and behavior will go on to influence others. To cite an example mentioned by Ormerod, can anything in conventional theory explain the runaway craze for the Teletubbies? Were millions of individuals acting independently and rationally in their enlightened self-interest? Or was there possibly instead an avalanche of interest that moved inexorably from one mind and one person to another? The forces of fashion are appreciable, and yet conventional economics does not even admit their existence.

The terrific success of some films, automobiles, books, hit records, and the like perhaps comes down to similar avalanches of interest. In 1994, the trade newspaper *Variety* compiled data on the gross theater revenues of the hundred most popular films of 1993. A plot of revenue versus rank (in terms of popularity) reveals a power law with broad tails, implying that the success of a film is extremely hard to predict.[21] The most popular film of 1993 earned more than forty times as much as that ranked number 100, even though this was still one of the year's hits. Why? Many of us know whether we want to see a film well *before* we go to see it, even before we have heard much about it. Somehow—through newspapers, television, word of mouth—we have come to an opinion. It is hard to deny that people get swept up in a craze and became interested when they know that others are interested. This is not rational decision making, but human beings influencing one another nonrationally.

In 1999, a pair of European researchers wondered if similar effects might have something to do with the terrific fluctuations that afflict financial markets. The economist Thomas Lux, of the University of Bonn in Germany, and the electrical engineer Michele Marchesi, of the University of Cagliari in Italy, set out to see whether the statistical properties of prices reflect external influences—changes in fundamentals as required by orthodox theory—or might instead be generated by "the mutual interactions of participants." To do so, they followed an approach that is by now familiar: they stripped down a stock market to a game of extreme simplicity, and used a computer to help them gain insight into its workings.

Imagine a stock exchange with only one kind of stock, and a population of traders who buy and sell it. In real life, the spectrum of strategies followed by traders is immense. Even so, Lux and Marchesi supposed that the mind-set of every trader places him or her at any moment into one of three broad groups. Fundamentalists are traders who stick to buying undervalued stocks—those with prices momentarily lower than their authentic value—and to selling overvalued stocks. In contrast to the Fundamentalists, Optimists believe that market prices are going up, and so want to buy stocks as wise investments. Pessimists believe that market prices are falling, and so want to sell stocks to cut their losses. The last two groups do not focus on fundamentals, but speculate on what they think are trends in the market.

The game works as follows. Lux and Marchesi assume that the stock has some true authentic value, determined by the fundamentals, and that this value fluctuates gently in the manner originally envisioned by Bachelier. The Fundamentalists keep a sharp eye on these fluctuations, and on the actual price, and buy and sell accordingly. The Optimists and Pessimists ignore the fundamentals and instead watch for trends in the actual price, which, of course, doesn't have to be the same as the stock's fundamental value. Finally, it is the interaction among traders in the market that works out the actual price. At any moment, there are a number of Fundamentalists, Optimists, and Pessimists all wanting to buy and sell stocks. The greater the demand for stocks, the higher the actual price; the greater the supply, the lower the price.

So far, all this fits in with orthodox economics, with a minor allowance for the very real fact that people do speculate on stock prices. But the key to the game comes in one further assumption: that people can affect one another.

Mind Games

Since human beings can influence one another, Lux and Marchesi assumed that the division of traders among the Fundamentalist, Optimist, and Pessimist mind-sets isn't fixed. Even people with strong convictions can be influenced by the actions of others, or by trends that seem simply too strong to ignore. A convinced Pessimist might, if market prices rise for a while, turn into an Optimist. A fanatical Fundamentalist might be persuaded by a long steady climb in prices that the market is experiencing a real and persistent trend, and that it would be silly not to cash in on it. Lux and Marchesi incorporate this into their game by supposing that at each moment, each trader has a small chance of changing his or her mind. If the Optimists outnumber the Pessimists, for example, then the prevailing opinion is that the market will continue to rise. Since people are affected by others' opinions, this makes it likely that more traders will soon become Optimists. If prices have been falling for some time, some Optimists may bail out and become Pessimists or Fundamentalists.

In essence, Lux and Marchesi use very simple rules to mimic how traders buy and sell stocks, and how their trading activity ultimately sets prices. They also include in these rules the possibility that people can change their strategies depending on what other people are doing. And this by itself, it turns out, is quite enough to drive the price of the stock on a wild roller-coaster ride. In running the game on a computer with a thousand traders, Lux and Marchesi insisted that the fundamentals fluctuate gently in keeping with the bell curve. These fluctuations drive ups and downs of a correspondingly gentle nature in the stock price. But out of the internal workings of the market there also emerge occasional far larger fluctuations, huge rallies or crashes, which appear to be triggered by absolutely nothing. Measuring the statistics of these fluctuations, Lux and Marchesi found them to match almost perfectly those of real markets, with self-similarity, structure on all time scales, and a distribution of price changes that looks just like the real thing: a power law revealing a great susceptibility to large fluctuations.[22]

There is little more at the core of the game than the possibility that one person can influence another. Yet on the basis of this feature, the network of traders organizes itself so that a tiny imbalance toward, say, optimism can lead to rising stock prices, and this rise makes optimism spread to still more traders, and makes the imbalance even greater. In response, the prices grow still higher, and so on in a self-sustaining chain of action and reaction. Ultimately the chain reaction comes to an end, and may then reverse itself. A few Fundamentalists may decide that stocks are so overvalued that they sell, triggering a small decrease in price. Suddenly, a few traders switch to being Pessimists, and prices drop still further. This downward surge may be momentary and small, or it may persist for a longer time, possibly even bringing prices back to their starting point.

All this is a long way from the conventional economic view in which corporate difficulties, political events, government decisions, and so on always lie behind any large market fluctuations. In the real world, traders speak of "runs" and "rallies," and refer to the market as having a "mood." In Lux and Marchesi's game, it really does, because all the traders within it have *their* moods. Since one mood can influence another, markets seem by nature always to be organized in something like a critical state, in which any momentary flicker of hope or doubt can be magnified out of all proportion.

The financier Bernard Baruch once suggested that

> all economic movements, by their very nature, are moti-
> vated by crowd psychology. Without due recognition of
> crowd-thinking . . . our theories of economics leave
> much to be desired. . . . It has always seemed to me that
> the periodic madnesses which afflict mankind must
> reflect some deeply rooted trait in human nature. . . . It is
> a force wholly impalpable . . . yet, knowledge of it is nec-
> essary to right judgement on passing events.[23]

In the light of Lux and Marchesi's model, Baruch's perspective certainly appears to be reflected in the mathematical reality of market fluctuations. Even if human beings are far more complicated than atomic magnets, grains of rice, or pieces of the Earth's crust, we are all similarly susceptible to influences, and, as a result, mass movements are common. The human world—at least in the context of the financial markets—seems to share the tumultuous and ever shifting character of the critical state. As a result, predicting the movement of a market may truly be impossible. A change in the mood of even a single investor may trigger a spreading wave of effects that leads to a fluctuation in the moods of nearly all investors.

What does this mean for the average investor? The news seems to be anything but comforting. I think most people know—or should know—that the ups and downs of the markets appear to be utterly unpredictable. Despite the confident predictions of bulls and bears, and despite what you may read in the newspapers, mathematical analysis indicates that no matter what the market has just done over the past week, month, or year, prices are still just as likely to go up in the near future as down. But this only hints at the true flagrancy of the market's unpredictability, or, shall we say, its upheavability. The power law for price fluctuations indicates that even the rough magnitude of the upcoming change is unforeseeable. In a market organized to the critical point, even the great stock market crashes are simply ordinary, expected events, although it is true that we should expect them infrequently. Even though there are no indicators whatsoever, the market may fall by 20 percent tomorrow. Such events need not be triggered by anything exceptional.

Can governments steer us away from such catastrophes? The idea would seem fairly unlikely, given that we cannot even see them coming. Nevertheless, economists have in recent years batted around at least one idea about how governments might put shackles on an economy and so reduce its upheavability by tuning it away from the critical point. The so-called Tobin tax, named after the economist James Tobin, would tax all speculative transactions—that is, all transactions deemed by some set of rules to be based on pure speculation about market trends, rather than on what was really happening to the fundamentals. The idea would be to dissuade investors from jumping on the bandwagon, thereby reducing the ability for influences to propagate. Would it work? No one really knows. Such a tax would, however, almost certainly have some deleterious effects on the market, such as reducing the overall trading volume; after all, the tax would penalize many of the transactions that go on now, and presumably some people would forgo those transactions in the face of a stiff tax. So would it be a good or bad idea to have a Tobin tax? As Lux sees it, "A serious economist should say, 'I don't know.'"

We may simply be stuck with these wild fluctuations. And there may even be something about the structure of social networks—in markets and elsewhere—that makes them even more naturally susceptible to upheaval than any of the physical systems we have considered so far.

Small Worlds

In 1967, an American social psychologist named Stanley Milgram conducted a peculiar experiment. Milgram sent a series of letters to various people in Kansas and Nebraska. Each letter, he explained, was intended ultimately for a certain stockbroker friend of his in the Boston area. Milgram didn't give the address of this friend, just his name and profession. Anyone receiving the letter was to speed it on its way by sending it to someone whom they knew personally, and whom they thought might have a better chance of knowing Milgram's friend the stockbroker. Almost miraculously, Milgram found that each letter reached its proper destination in about six steps. That is, in just six steps it found someone who actually knew the Boston stockbroker, and who then sent it to him directly.

So was born the idea of "six degrees of separation," a now-popular notion that seems as unlikely as it is appealing. There are more than 6 billion people inhabiting our planet. Nevertheless, the claim goes, every one of us is connected to every other by a chain of linked acquaintances involving no more than six people. As a character in John Guare's 1990 play *Six Degrees of Separation* puts it:

Everybody on this planet is separated by only six other
people. . . . The president of the United States. A gondo-
lier in Venice. . . . It's not just the big names. It's anyone.
A native in a rain forest. A Tierra del Fuegan. An Eskimo.
I am bound to everyone on this planet by a trail of six
people. It's a profound thought. . . .

If the idea is true, it would certainly go a long way toward explaining
how it is that you travel to Thailand or Alaska, or telephone Zambia,
and somehow meet or find yourself talking with someone who knows
the wife of your former thesis advisor, or the father of your best
friend, or your mother-in-law's hairdresser. Are these really stagger-
ingly improbable coincidences? If so, why do they happen so frequently? Or is Milgram right and the world really is small, at least in
human terms? In 1998, the mathematicians Duncan Watts and Steve
Strogatz of Cornell University waded into the theory of graphs in an
effort to sort things out.

To a mathematician, a graph is a grid of dots with some lines
linking the dots. It is a good way to represent people (the dots) and
the connections they have to one another (the lines). You wouldn't
expect that you could learn much from dots and lines, but you can.
One extreme in the world of possible graphs is the random graph,
formed by choosing pairs of dots at random and drawing lines
between them. The result is a tangled spaghetti-like diagram. At the
other extreme is the ordered graph, with highly regular links connecting dots that are close to one another, so that the graph ends up
looking like a fishing net or a fence (SEE FIGURE 17, A AND B).

Getting around on a random graph is very easy. If you start at a
dot on one side, for example, it takes only a few steps to reach any
dot on the other side. This is because there will almost certainly be
some long-distance link with its ends fairly close to these two dots,
and by taking this "shortcut" you can go from one to the other
quickly. Ordered graphs, by contrast, do not have this small-world

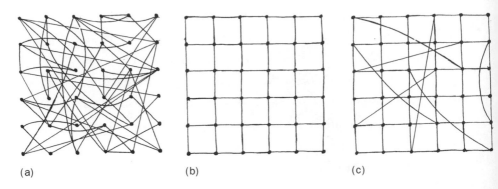

(a) (b) (c)

FIGURE 17. You can get anywhere in just a few steps on a random graph (*a*), while it takes many more steps to move about on an ordered graph (*b*). Small-world graphs (*c*) live between the two extremes.

property, since there aren't any shortcuts. You have to take short steps all the way. Are real-world social networks like random graphs? If so, this might explain how Milgram's letters got there so quickly. But there is a serious stumbling block.

Consider your own group of friends. Many of them are friendly not only with you, but also with one another. This is natural and typical of networks of friends. On the social graph, many of the dots for your friends should be connected not only to you, but also to one another. Ordered graphs have this clustering property, but random graphs do not. Take one dot on a random graph and consider all dots connected to it by a single link. These will be scattered all over the place, and will only rarely be linked to one another. If a social network were like a random graph, then there would be no such thing as a group of friends.

Through their small-world character, then, random graphs seem to mimic the incredible connectedness of real social networks. But it is ordered graphs that show the clustering typical of such networks. This would be a paradox, were there not another class of graphs between the two extremes. Consider an ordered graph on which you do a bit of "rewiring" by breaking a few of the short links and replacing them with a few random long-distance links (SEE FIGURE 17, C). Watts and Strogatz studied the effects of such changes and discovered that these few shortcuts have very little effect on the clustering within the graph. Nevertheless, they have a tremendous effect on the average number of steps it takes to get from any one dot to another. That is, only a very few shortcuts can change an orderly graph into a small-world graph that still has clustering, but on which just a few steps will suffice to take you just about anywhere.[24]

To see if real social networks might really be organized in this way, Watts and Strogatz turned, oddly enough, to the world of acting. Finding good data on networks of friends is not so easy. But who played in what film for more than half a century is all there for inspection. Imagine a graph with one dot for each actor and links connecting actors who have ever played together in some film. According to lore, very few actors who have ever played in an American film stand more than about four steps away on this graph from the actor Kevin Bacon—notorious for being in lots of films, though not often being the star. Elvis Presley is only two steps away from Bacon, since he played with Walter Matthau in *King Creole* (1958), and Matthau played alongside Bacon in *JFK* (1991).

But Watts and Strogatz found that Bacon is not really all that special, because the entire network has the small-world property. Every actor is linked to every other by typically just three or four steps. Although the numbers may change slightly, all social networks presumably have this same small-world character, which appears to be the mathematical secret of how each of us is linked to anyone from Monica Lewinsky to Boris Yeltsin by just "six handshakes." But there may be a deeper message here as well.

The small-world character of real social networks allowed Milgram's letters to move quickly. Watts and Strogatz also modeled the spread of infectious diseases on small-world networks, and found that they spread much faster than they would on ordered networks.

What's more, only a very few shortcut links are necessary to make this happen. This has disturbing implications for how dangerous diseases might be able to spread over the world, carried to or from remote places by just a few long-distance travelers.

And what about the spread of ideas? As we have seen in this chapter, financial markets are wild at heart, seemingly because the opinions or expectations of one investor can affect those of another. The small-world character of the social and business relationships among traders in financial markets simply adds to the ease with which such influences can be amplified immensely. It may, in fact, make it even easier for the wild fluctuations of the critical state to rip through the markets. As the notion of the small-world network has only just been discovered, its ultimate consequences for the workings of social networks of all kinds remain to be seen.

· 9 ·

Against All Will

The capacity to get free is nothing;
the capacity to be free, that is the task.

—ANDRÉ GIDE[1]

———————

Whether it is a good thing or a bad thing,
smashing things up is sometimes very pleasant.

—FYODOR DOSTOYEVSKY[2]

AN INSECT OF DOUBT MAY BE CREEPING INTO THE MINDS OF SOME
readers. Isn't there a contradiction lurking in the ideas of the previous chapter? What about that most precious of all human posses-
sions, our free will? As I write, I believe I choose my own words and
could choose them otherwise. You believe that you picked up this
particular book by choice, and that you could have selected another.
Similarly, traders and investors on Wall Street are independent, free-
thinking people, and can on any day choose to buy or sell any of a
thousand different stocks, bonds, or options, or instead to do nothing at all. People are not grains of rice or sand that tumble according to preordained rules.

Once you become accustomed to the idea, it is not so hard to
accept that the essential logic of the critical state might arise in simple physical things such as a pile of sand, or even in the rocks of the
Earth's crust or the trees of the forest, where definite physical laws
control how activity spreads from one place to another. When the
stress becomes too great somewhere along a fault, the rocks slip,
shifting stress onto rocks farther down the line. In cases such as this,
there is no need to reckon with anything so ineffable and capricious
as a thought or an emotion. But when people get involved, things
aren't so simple. People decide on their own whether influences will
propagate or not. So despite the mathematical evidence, aren't we
making a dangerous leap in supposing that the critical state has anything to do with the human world?

Beginning with chapter 10, the final four chapters of this book
will look at the critical state in the context of science and human history, and consider whether it might shed some light on the origins of
tumultuous events in those settings. So we need briefly to examine
this issue of free will, and see whether it really does set up a kind of

inviolable barrier against the intrusion of mathematical regularities into the human world. As we shall see, the answer is clearly, even obviously no. We all accept quite easily that marriage rates in Britain can be falling slowly and steadily even though every individual person continues to make up his or her own mind on whether to marry. So there is clearly nothing in individual free will that would preclude the emergence of definite mathematical patterns in the actions of thousands or millions of people. There are better ways, however, to make the point.

Trail Finders

Set in among the imposing stone and brick buildings of any university campus, you will find quiet, empty spaces—squares and commons, lawns and grassy retreats, persistent reminders of the open-minded atmosphere conducive to learning. Students gather in such places to sit in the sunlight or to eat lunch, to sleep, or read, or just to think. But if they wanted to, they could also quite easily take a lesson in the mathematics of human behavior. The architects of such areas typically put in cement walkways with straight lines and sharp angles to help people move about, and, just as typically, the ever rebellious students follow their own routes instead, carving out over time a series of well-worn and meandering dirt trails.

These trails range from the obvious shortcuts across places where no walkway leads to far more exotic networks of mingling trails, such as you can find at the University of Stuttgart in Germany (SEE FIGURE 18). Physicist Dirk Helbing walks on these particular trails frequently, and in 1996 he began wondering: Might it be possible to work out the laws for their evolution? Can the formation of trails be predicted? In crossing such an area, of course, each student chooses freely, and none need follow another. Even so, as Helbing and his

FIGURE 18. A human trail system that has evolved on the campus of the University of Stuttgart. Photograph courtesy of Dirk Helbing, University of Stuttgart, reprinted by permission.

colleagues soon discovered, this doesn't preclude trails from growing in accordance with laws that are every bit as definite as those for the motions of the planets.

Why trails form at all is easy to see, for while individuals may indeed follow the demands of their precious free will, they also have tendencies. Imagine a grassy common or square long before any trails have formed. In crossing, people obviously head in the direction they want to go—to reach a pub on the far side, perhaps, or a class farther on. But rarely does anyone follow a perfectly straight line. People also prefer to walk where it is easiest, skipping around puddles, avoiding rough or muddy areas or the wet grass. They generally follow the smooth cement. But staying on the cement isn't always satisfactory either, for in following it to the pub across the way, the person may take an annoying detour around the entire edge of the square. If so, then anyone who is thirsty faces a dilemma.

Not everyone is going to the pub, of course, but others face similar decisions about whether to stay on cement or plunge into the pristine grass. At first there is nothing more to say; each person simply makes up his or her own mind. And yet as people choose and the footsteps rain down, things begin to change. When a person plunges into the grass and tramples it down, the particular route he or she follows becomes a tiny bit more attractive to future walkers. The difference may be minuscule after one person, but after a thousand a new trail begins to emerge, and the trail then lures still more walkers to abandon the cement. At some point the path becomes so well worn that people plod over it with the mindlessness of cattle, and it remains there for good.

This is the rough story behind trail formation. But Helbing and his colleagues found that this story actually forms the basis of a thoroughgoing theory. It takes only a few simple equations to describe how each walker, on average, tries to strike a balance between shorter and easier paths, how the grass gets trampled as people walk over it, and how this wears down the turf, altering the layout of paths available to future walkers. To use the equations, you specify the geometry of the grassy area in question, the locations of the most popular destinations, how many people enter each day from each side, and so on. Using a computer, you then send a few tens of thousands walking across the green and see where the trails emerge.

They emerge, it turns out, more or less where they do in the real world. For conditions like those for the green at Stuttgart, the researchers found that the equations spell out an elegant system wherein three trails run toward the center of the green and there intersect with a central triangle of shorter trails, much as in the real case. Not only do the equations lead to this pattern, they also offer some insight into why it forms. A limited number of people walk across any green space each day. So the total length of the trails it can support is also limited: the people can manage only so much trampling, and grass grows back on any underused trails. Within the context of this constraint, it turns out that the trails that form produce an "optimal" system—that is, the layout enables people, insofar as possible, to get about on paths that are both fairly short and easy to walk on.

This theoretical demonstration is only a beginning. By using these equations, planners ought to be able to choose the sizes and shapes of green spaces and to place buildings and cement walks where they will complement human tendencies. Doing so will require the gathering of crucial practical information, such as the locations of the most popular destinations in the area and the number of people who will be crossing the space each day. In any event, it now seems indisputable that human trails do form according to the very specific laws laid down in Helbing and his colleagues' equations.[3]

This example has very little to do with the critical state, and I have brought it up only to illustrate how easily individual free will can coexist with stunning regularity in the activity of a group. The peculiar network at the University of Stuttgart emerged from the free actions of thousands of people all following their own intentions, and yet it conforms to very simple mathematical rules.

Metro-mechanisms

To seek what has made one city large and another small is to wrestle with myriad social and economic forces, as well as innumerable historical and geographical facts. During the American Civil War, the Confederacy established its capital in Richmond, Virginia. Richmond today is a city of nine hundred thousand, and yet it might easily be five times that large if the Confederacy had managed to establish its independence. Washington, D.C., sprouted up as the capital of a nation, Chicago emerged as a key link between the eastern and western United States, and cities such as Pittsburgh and Cleveland in the American Midwest grew as mighty centers for the steel industry. In contrast, Charlottesville, Virginia, still relatively untouched by major industry of any kind, has remained small, despite its proximity to Washington.

In view of the tangle of forces affecting how any city grows, and also in the knowledge that people move from one place to another for their own very personal reasons, we might well despair of finding any kind of mathematical regularity in a study of the cities. In 1997, however, Damián Zanette and Susanna Manrubia of the Fritz Haber Institute in Berlin, Germany, found otherwise. Forget all those details concerning the histories of cities such as Chicago or Cleveland: the pattern is to be found by looking at all the cities at once.

Using data for the twenty-four hundred largest cities in the United States, Zanette and Manrubia counted up how many cities there are with populations of around one hundred thousand, two hundred thousand, three hundred thousand, and so on, all the way up to cities with populations of about nine million, of which there is only New York. In other words, they approached cities just as Gutenberg and Richter did earthquakes. And they found a similar pattern. The numbers reveal that for every city such as Atlanta, Georgia, population four million, there are four cities having populations half that size. Cincinnati, Ohio, is one such city, and for every Cincinnati there are in turn four cities just half as large again, and so on down the line. This perfect geometrical pattern continues down to the smallest cities of just ten thousand people or so. So even while all these cities have sprung up for a thousand reasons and as the result of a million competing influences, they nevertheless conform in their totality to a simple mathematical law.

Given the free choices that shuttle people between one city and another, a pattern of such striking regularity may be surprising.

Zanette and Manrubia didn't stop with cities in the United States, but looked also at the twenty-seven hundred largest cities worldwide and at the thirteen hundred largest communities of Switzerland. And in every case they found precisely the same power-law pattern, which seems to be a universal consequence of the process by which people aggregate into cities. This is especially striking, Zanette and Manrubia point out, as the same pattern emerges

> . . . in spite of the fact that the three data sets correspond to very different demographic, social and economic conditions. In fact, the data for the world is expected mainly to reflect the situation of developing countries, the United States is an economically developed but young nation, whereas Switzerland is an old country with a relatively very stable population.[4]

In other words, here is a kind of universality at work at the level of people. Somehow, all these differences influence not at all the relative numbers of larger, smaller, and medium-size cities that grow up.

The implications of this power-law pattern are as usual: there is no "typical" size for a city in the United States or elsewhere, and no reason to see special historical or geographical situations behind the emergence of the very biggest. The growth of a city is a critical process much like those we have seen already, poised on the edge of great instability. You might imagine that when a town is founded, because of its location, its industry, and other factors, it might be destined for greatness. But the power law suggests that there is no telling at the outset how big a town will become. There was probably nothing inevitable or special about the beginnings of New York, Mexico City, or Tokyo. If you could rewind history and play it again there would no doubt be great cities, but in different locations and with other names. Even so, the power-law pattern of cities would remain the same.

So there can be a mathematics for people. It cannot tell you, of course, what any one person will do, and yet it may be able to say what kinds of patterns are likely to emerge out of the millions. What's more, the mathematics is not complicated. Zanette and Manrubia were able to capture the essence of city growth in an outrageously simple game. Since human decisions to move, to have children, and so on are extremely unpredictable, they supposed that the changes in population in any place from year to year are more or less random—with one qualification. It makes sense to expect the raw numbers for the fluctuations in population of any large place such as New York City to be higher than for those of a smaller place such as Lubbock, Texas. Zanette and Manrubia built this into their game, supposing that these raw numbers for the yearly changes in any place should grow in direct proportion to the number already there. In other words, the more people, the more vigorously the population fluctuates. *"mass action"*

It is also true that people tend to "flow" from places of high population toward those with low, as they seek more space, cheaper property, or what have you. This latter influence should tend to smooth the population out, obliterating the cities and spreading a uniform film of humanity over the land. But in a setting no more complicated than the sandpile game, Zanette and Manrubia discovered that this smoothing influence isn't a match for the fluctuations. The fluctuations keep stirring up population differences in a way that leads inevitably to the growth of "cities"—clusters of aggregated people—and reproduces the power law for city sizes. So you can forget all those economic factors and geographical constraints. The process by which cities grow all over the world is in some respects a far simpler thing than you might think.

The simplicity extends even further, for there are also definite patterns regarding where people live within any particular city. Aerial photos at night over London or Berlin would reveal very different images—different, that is, in their precise details. But under closer inspection, such images turn out to be strikingly similar. There are larger and smaller clusters of population sprinkled within any city, and such clusters follow a power law. So there is no "typical" size for a cluster, and there is a kind of self-similarity in the overall settlement of people: any small cluster when blown up looks again much like the whole, with smaller clusters within it.

So while all cities are different, they are also deeply alike. Cities are fractals—self-similar patterns—just like our two-dimensional magnet at its critical point. And ironically, but perhaps not surprisingly, the best way currently known to describe the pattern of population within any city is by using simple games from the theory of phase transitions.[5]

Rags to Riches

One of the messages of universality is that understanding something often means looking past the surface details to spy the deeper logic beneath. As we have just seen, the way people aggregate into cities doesn't depend at all on the fact that they are people. It may be somewhat insulting to say this, but similar patterns can show up for similar reasons in mindless colonies of aggregating bacteria or clusters of smoke particles deposited on a ceiling. And they also show up in another place: in the way money gathers—or, all too frequently, fails to gather—in our pockets and bank accounts.

Why does one person become rich and another poor? As with the cities, the reasons are many, and any answer certainly has to refer to the situation into which a person was born, his or her chances for education, and the like. But despite all the advantages, disadvantages, and differences in individual ability, there is a simple pattern. If you tally up how many people in the United States have a net worth of a billion dollars, you will find that about four times as many have net worth of about half a billion. Four times as many again are worth a quarter of a billion, and so on. If this special pattern held for just one country under one government at one point in time, then perhaps you might write it off as a peculiar quirk of some governmental policy. But the very same pattern holds in Britain, the United States, Japan, and virtually every country on Earth.

Earlier this year the French physicists Marc Mézard and Jean Philippe Bouchaud were able to explain this pattern in a way not too different from that of Zannette and Manrubia. Suppose that each person's wealth grows or shrinks by a random fraction each year. There are no "sure thing" investments, and so the income a person generates is truly random in any year. But the size of these random changes should be proportional to the person's wealth, since richer people can invest more, and so can either gain or lose more than poorer. Suppose also that each person contributes to the wealth of some other people by virtue of working for them, investing money in their business, and so on. It is hard to argue with these very basic assumptions. And yet Mézard and Bouchard found that in a simple game including only these effects, the power-law distribution of wealth comes tumbling out.[6]

So again, even though people interact with one another by virtue of their own personal decisions, suspicions, plans, and schemes, there nevertheless emerges a very regular pattern. And that pattern seems to have far less to do with the nature of people as people than with a universal kind of organization that tends to well up in any collection of interacting things. This way of thinking does not let anyone predict who will get rich and who will not. But it does begin to explain what we might call the basic physics of money flow and aggregation.

By this point, I hope, it should be obvious that there is no reason to suppose that the individual human will get in the way of mathematical laws operating at the level of large groups of people. To suppose that it would is merely to neglect the distinction between laws for individual objects—atoms, people, or what have you—and laws for collections of large numbers of such things. In physics there are lawlike regularities in both cases: the magnets inside a piece of iron flip one way or another according to definite and simple physical laws, and the interactions between huge numbers of them lead to equally regular laws for the iron itself. In the human world, there may be no laws for the individuals, yet this needn't imply a similar lawlessness for the many.

Some people might wonder if these power laws couldn't be explained in some other way. Just because the same pattern turns up in different places, this doesn't necessarily imply that the same cause is at work. If all the trees on your property fall over one night, you might suppose that one fell over because its roots had decayed away, that another was pulled over by your meddling neighbor, and that there were other distinct causes for all of the other trees. Or you might instead seek a simpler explanation. Remembering that there was a terrific windstorm last night, and that the trees all happened to fall in the same direction, you might guess that the storm brought them all down.

Similarly, the simplest explanation by far for the appearance of all these spectacularly simple power laws is that some universal process is at work. This idea becomes even more appealing in view of the universality that we know is at work in systems that are made of many interacting elements, and that operate in the same way regardless of almost every last detail of those elements. What's more, outside of the realm of non-equilibrium physics, there are very few tricks that can produce power laws.

· 10 ·

Intellectual Earthquakes

History cannot create laws with predictive power.
An understanding of the past might help in the
present insofar as it broadens our knowledge of human
nature, provides us with inspiration or a warning,
or suggests plausible, though always fallible arguments
about the likely possibilities of certain things
happening under certain conditions.
None of this, however, comes anywhere near the
immutable predictive certainty of a scientific law.

—RICHARD EVANS[1]

Eventually, everything we currently believe will be revised.
What we believe, then, is necessarily untrue. We can
only believe in things that are not the truth . . . I think.

—MAX GUYLL

◆

WHAT *REALLY* CAUSED THE FIRST WORLD WAR? IF THE SERBIAN TERRORIST
Gavrilo Princip obviously sparked things off outwardly, what were
the deeper forces that precipitated what many then viewed as "the
greatest calamity that has ever befallen the human race"?[2] Historians
just after the war had more than a few ideas. In America, the historian Sidney Fay pointed to flaws in the international system, including a tangled web of hidden military assurances and insufficient
political means to settle disputes.[3] And in Russia, not surprisingly,
the Bolsheviks attributed the war to a kind of natural meltdown in
the capitalistic world. Many other historians saw simple German
treachery as the real cause. The American historian Charles Beard
poked fun at the simple-mindedness of this prevailing view, which he
called the "Sunday-school theory":

> . . . three pure and innocent boys—Russia, France and
> England—without military guile in their hearts, were
> suddenly assailed while on their way to Sunday school by
> two deep-dyed villains—Germany and Austria—who had
> long been plotting cruel deeds in the dark.[4]

Subsequent historians have come to agree with Beard that maybe
this was all a bit too simple, even if they have not adopted the contrasting perspective of Beard's contemporary, Harry Elmer Barnes.
As if to demonstrate just how wildly serious historians' views can dif-
fer, Barnes concluded that

> the only direct and immediate responsibility for the
> World War falls upon France and Russia, with the guilt
> about equally distributed. Next in order—far below

France and Russia—would come Austria, though she never desired a general European war. Finally, we should place Germany and England tied for last place, both being opposed to war in the 1914 crisis. Probably the German public was somewhat more favourable to military activities than the English people, but . . . the Kaiser made much more strenuous efforts to preserve the peace of Europe in 1914 than did Sir Edward Grey.[5]

Today, there is still no general agreement on the war's ultimate causes. Nor have historians reached any final and definitive consensus on the causes of many other events, ranging from the American Civil War to the Norman Conquest of 1066. Not that this is very surprising. There are, after all, no deterministic laws of history, no historical equations or even deep and fundamental principles to which researchers can appeal in trying to account for this or that event. In the law of gravitation, physicists find the principal explanation for the motion of the planets and the shapes of galaxies. But history is not like physics. In history, frozen accidents continually alter the playing field on which the future must unfold, and so the historian can only fall back on the telling of stories.

To explain why Eisenhower's allied armies were standing along the Rhine in the fall of 1944, we have to refer to the First World War and its ignominious ending for Germany, to Hitler's rise to power in 1933, the German army's successes in France and the rest of western Europe, and its ultimate defeat in Russia. We cannot ignore the American Lend-Lease program that was so important in supplying war materials to both Britain and Russia, or the Japanese attack at Pearl Harbor that brought the United States into the war. We also have to take account of myriad events on the battlefield, such as the fateful order from Hitler on May 24, 1940, that halted the decisive advance of General Heinz Guderian's First Panzer Division when it was just ten miles away from Dunkirk. Had Hitler kept his mouth shut, Guderian's armored columns would have captured or destroyed the entire British Expeditionary Force. Alter just one of these or a thousand other facts, and Eisenhower's forces might never have stood along the Rhine.

But which of these events were the more crucial, and which the less? Here the historian's personal taste comes into play. Some seek the true causes of important events in stories of political intrigue; others look instead to the interplay of economic, sociological, or cultural forces; and others look to the decisive personal influence on events of a Hitler or a Stalin. So even having lived through the same events, and having access to the same documents, historians still tell different stories.

This is one of the unavoidable problems with which historians grapple. For the sake of argument, however, suppose that all historians could agree—that in considering any event, every historian, after sufficient study, would come to tell exactly the same story. What would this story *really* explain? Would it capture everything there is to explain about a dramatic event such as the First World War? And what, if anything, would such a story leave unexplained? We will forget real history for the moment, and put storytelling to the test in a far simpler historical setting.

Sandman His Own Historian

Imagine a historian of the sandpile world who one day finds his community swept up in an enormous avalanche. How does he (or she) explain what happened? After some investigation, the historian might offer a report like the following:

> The trouble began a week ago in the remote West, where in the early evening a single grain of sand fell on a portion of our pile that was already very steep. This triggered a small avalanche, as a few grains toppled downhill

toward the East. Unfortunately, the pile hasn't been managed properly in the West, and these few grains entered into another region of the pile that was also already steep. Soon more grains toppled and throughout the night the avalanche grew in size; by the next morning, it was well out of control. In retrospect, there is nothing surprising. One fateful grain falling a week ago led to a chain of events that swept catastrophe across the pile and into our own backyard here in the East. Had the Western authorities been more responsible, they could have removed some sand from the initial spot, and then none of this would have happened. It is a tragedy that we can only hope will never be repeated.

This story of what happened is undoubtedly of compelling interest to the historian and everyone else affected by the disaster. But does it explain anything about why the catastrophe happened? In the sandpile, every avalanche large or small can be "explained" by giving a grain-by-grain account of the action. This proves that the grains obeyed the laws of granular physics. But there is a deeper question: What made it possible for a single grain to trigger a pile-wide catastrophe?

The sandpile historian imagines that he has put his finger on the special conditions in the West that made the catastrophe happen. "If only someone had acted sooner," he cries, "and removed some sand from that initial site!" Yet this is at best a comforting illusion. No amount of prior investigation carried out near the place where the grain was about to fall would have turned up any unusual "precursory" details. If the pile was steep there, it was also steep at many other sites all across the pile, at any of which a falling grain would have triggered nothing remarkable. To have foreseen the disaster, our historian would have needed near-perfect knowledge of the placement of grains over the entire pile, coupled with virtually unlimited computing power to work out the consequences of a grain falling at any possible position. Only then would it have been possible to say with confidence, "Yes, it is certain, if a single grain falls at danger spot X in the West, there will be an immense catastrophe."

What's more, while it may be true that removing just a single grain from that initial site would have prevented the disaster, there is no way to know beforehand which grains should be moved and to where. Had the Western authorities moved a few grains, they might have been dismayed a few weeks later. For they might have discovered that a grain falling somewhere else on the pile had caused an avalanche that became catastrophic just because of the grains that they had moved. In that case, the historian would blame the West for causing a disaster, rather than for failing to prevent one.

Unfortunately for the historian, his narrative can sketch out only specific chains of events, and cannot touch the deeper historical process that lies behind them. With his narrative, he can pay homage to the quirky contingency of history and be done with it. But ask him why all avalanches can't be small, and he cannot answer. To understand why a single grain can trigger a cataclysm, one needs to understand the detailed structure of the pile not just in a small region, but over its entirety. And one needs to understand about those long fingers of instability that run through the pile. Only in this way can the historian gain a far deeper appreciation of history—an appreciation not only of *what* happened, but of *why* something of that general character had to happen, and will undoubtedly happen again.

When it comes to human history, of course, no historian is forced to stick to the bare narrative. But how can he or she get hold of something deeper?

More Than a Story

It was the great nineteenth-century German historian Leopold von Ranke who originally enthroned the narrative when he identified the historian's task as *wie es eigentlich gewesen:* merely "to say how it essentially was."[6] For some historians, this hasn't been enough. Forty years ago, the Oxford historian Edward Hallett Carr lamented how

> three generations of German, British, and even French historians marched into battle intoning the magic words *"wie es eigentlich gewesen"* like an incantation designed, like most incantations, to save them from the tiresome obligation to think for themselves.[7]

In Carr's view, it wasn't the mere telling of specific tales, but the generalization from them, that was the real point of doing history:

> The very use of language commits the historian, like the scientist, to generalisation. The Peloponnesian War and the Second World War were very different, and both were unique. But the historian calls them both wars, and only the pedant will protest. When Gibbon wrote of both the establishment of Christianity by Constantine and the rise of Islam as Revolutions, he was generalising two unique events. Modern historians do the same when they write of the English, French, Russian and Chinese revolutions. The historian is not really interested in the unique, but in what is general in the unique.[8]

So what is "general in the unique"? What are the generalizations of history? There are undoubtedly many to which historians might point, but one of the most obvious and fundamental was highlighted more than half a century ago by the American historian Conyers Read. One of the important lessons available from the study of history, Read suggested, is that

> unless we are alert to the necessity of constant re-adjustment we create a condition of maladjustment

which is the inevitable forerunner of Revolution, whether that Revolution take the Russian or the Italian form. . . . I believe that the study of history has an important social function to perform of just this sort.[9]

In other words, Read's "maladjustment"—a building up of some kind of internal stress—precedes any episode of revolutionary upheaval. Or, as Thomas Carlyle commented with regard to the origin of the French Revolution:

Hunger and nakedness and nightmare oppression lying heavy on twenty-five million hearts: this, not the wounded vanities or contradicted philosophies of philosophical advocates, rich shopkeepers, rural noblesse, was the prime mover in the French revolution; as the like will be in all such revolutions, in all countries.[10]

Maladjustment, according to historians, is the precondition of revolution, and necessarily precedes all sudden and dramatic changes, in all communities, no matter what their character or size.[11] Implicit in this observation also is the recognition that the maladjustment, and the human distress that goes with it, have to reach some kind of threshold of severity before the social fabric will give way. The distress, in other words, has to be great enough to overcome what another historian calls "that greatest of all social forces—inertia."[12] Revolutions certainly do not appear to be happening every day, even though some elements of all societies must always be dissatisfied with the existing order of things.

This generalization of history may seem so obvious and vague as to be either meaningless or simply true by definition. But it is both suggestive and intriguing to compare the idea with the basic physics of the sandpile. There, an avalanche starts only when the slope at some point becomes so steep that the next falling grain pushes it past a threshold, and sand begins sliding. Similarly, in the Earth's crust, the stress of "maladjustment" builds up in the rocks until finally and suddenly they give way in an earthquake. If the generalization noted by Read is truly general, then it is not wholly frivolous to suppose that revolutions, wars, and other dramatic social upheavals may all reflect the workings of an underlying historical process with the same susceptibility to upheaval that we have seen so many times already.

We shall return to this possibility in the next chapter. But before we swim out into the raging river of all human history, it will help to wade briefly into one of its narrower streams: the history of science. If human history has a general character, then it should be visible in any of its particular aspects. Nearly four decades ago the historian of science Thomas Kuhn published a remarkable book that in one powerful stroke upset most of prevailing ideas about how scientists do their work. As we shall see, Kuhn very clearly identified science as one setting in which this universal building up and release of stress has a telling influence on the tempo and character of history. So as a step toward understanding what might lie behind wars and political revolutions, it is salutary to look at what lies behind scientific revolutions.

Habits of Learning

At the end of the nineteenth century, science still lived in the Age of Innocence. Scientists were widely considered almost superhuman in their ability to be open-minded, rational, and objective, and to follow the infallible principles of the scientific method. A view then common held that scientists formulate hypotheses about how things might be, test them against objective reality, and retain only those ideas that "fit the facts." Any idea that doesn't measure up, they simply toss out like a wad of old chewing gum, with no regrets whatever.

Science *is*, of course, about inventing and testing ideas, and coming to beliefs through conversation with nature; it is decidedly not about being told "how it is" by some authority. "Science," as Richard Feynman once expressed it, "is belief in the ignorance of experts"—and, one might add, in the possibility of becoming slightly less ignorant through careful investigation. But while this is true, there is nevertheless a great naïveté in any view that would see the scientist as some kind of automaton driven by the Holy Trinity of Rationality, Objectivity, and Open-mindedness. Scientists are human beings, and since all science takes place in the setting of a community of researchers, scientists can influence other scientists. In the 1950s, a number of historians began to see that this simple possibility has important consequences.

On the basis of detailed historical studies of how science has really worked in practice, for example, the historian Michael Polanyi came to the conclusion that scientists are not actually so open-minded and rational as they might have you believe. Instead, he found that

> there must be at all times a predominant accepted scientific view of the nature of things. . . . A strong presumption . . . must prevail . . . that any evidence which contradicts this view is invalid. Such evidence has to be disregarded, even if it cannot be accounted for, in the hope that it will turn out to be false or irrelevant.[13]

Instead of always being open-minded, Polanyi found, scientists often have their minds and eyes closed. Rather than always seeking evidence to test their ideas, they often ignore such evidence even when it hits them in the face.

At Harvard University, Kuhn produced several long historical studies of some of the most dramatic episodes in science, such as the Copernican revolution, and the upheavals associated with the births of relativity and quantum theory. In each case, he similarly found that scientists did not promptly reject the old theories when these were judged rationally and objectively, in the courtroom of facts, to be lacking. Kuhn noticed instead that scientists at any moment seem to be emotionally committed to a shared set of ideas, and will not even consider rejecting these ideas unless their "maladjustment" to the nature they are meant to describe becomes obviously and intolerably great.

In retrospect, none of this should be surprising. After all, scientists aren't superhuman, and, when doing science, they aren't even all that different from anyone else. They suffer the usual human biases and blindnesses, and often "want" the world to turn out to be one way rather than another. None of this implies that science does not work. Indeed, it seems to work, and spectacularly well. But how does it work? And how does it manage to grow if scientists often refuse to relinquish their cherished ideas? As a historian who shared Carr's desire to find generalizations in history, Kuhn set out to answer these questions, and did so in his classic 1962 book *The Structure of Scientific Revolutions.*

At the structural center of his historically more realistic image of science, Kuhn located the notion of a paradigm. A paradigm is any concrete example of some scientific ideas or practices that have been proven to work. Paradigms are, in Kuhn's words,

> . . . accepted examples of actual scientific practice—examples which include law, theory, application, and instrumentation together—providing models from which spring particular coherent traditions of scientific research.[14]

Put Newton's equations together with their mathematical applications to planetary motion, and you have a paradigm. As another, consider Maxwell's equations for electricity and magnetism, coupled with the practical rules for applying them to the workings of radio waves, electrical generators, and so on. The principles and practices of the quantum theory represent another paradigm, one that is now relied upon daily by thousands of physicists. Think of a paradigm as a "bundle of Good Ideas" that allows scientists to explain some things that have always been mysterious.[15]

Without a paradigm, a scientist would drown in the bewildering sea of natural phenomena, unable to tell which facts are important and which aren't. In their training, scientists learn a variety of paradigms, and so learn by example how to do science. These bundles of ideas tell scientists what kinds of things the universe is made of—atoms, waves, quantum fields, or what have you—and specify the basics of how these things behave. As a result, they make the doing of science largely mechanical. The "Good Ideas" of a paradigm give the scientist his or her foundations, and, consequently, scientists commit themselves to a paradigm with considerable fervor.

The collection of all scientific paradigms, then, forms a kind of network of Good Ideas glued together and fixed in place by scientists' collective commitment to it. The most obvious paradigms are the most basic Good Ideas: quantum theory, relativity, the theory of evolution, and the like. But there are countless smaller Good Ideas that fill in the network as well, ideas that have proved themselves somewhere along the line, and instruct scientists on how to solve certain kinds of equations, or what kinds of experimental procedures give good results. All these ideas together form the core structure at the heart of science, the "accepted scientific view of the nature of things" to which Polanyi had referred.

But the main project of science is to learn more—that is, to make the network of good ideas more dense and complete. If science is about learning, then this network can hardly remain fixed, and Kuhn identified two basic but essentially contrasting ways by which it can change.

What's Normal and What Isn't

Even if a bundle of ideas makes the world in certain aspects seem reasonable, there is considerable work to be done to discover just exactly what these ideas imply. Many physicists, for instance, are currently trying to solve the puzzle of sono-luminescence, a bizarre but easily demonstrable happening in which focused sound can make water glow brilliantly. Even though the puzzle has been around for decades, everyone assumes—and they are probably right—that a combination of chemistry, quantum theory, and the physics of fluids is fully capable of making sense of whatever it is that is going on. The project, in other words, is to take ideas to which scientists are already committed and to reveal how they make sense of an increasing slice of the world.

Kuhn called this "normal science." It is the activity that aims to elaborate the paradigm, to work out everything that its ideas imply. It might well be likened to a kind of simple growth. This kind of science is very conservative, since it does not question the good ideas of any paradigm, but takes it on faith that the accepted view of the nature of things holds the key to understanding almost everything. As Kuhn observed,

> Normal science . . . whether historically or in the contemporary laboratory . . . seems an attempt to force nature into the preformed and relatively inflexible box that the paradigm supplies. No part of the aim of normal science is to call forth new sorts of phenomena; indeed those that will not fit the box are often not seen at all.[16]

Normal scientific work is the activity that aims to extend the network of Good Ideas so that it covers a greater portion of nature, to fill in any gaps, and generally to make it a complete and seamless whole.

But not all science is normal. Efforts to make the network grow in one direction or another, or to fill in some empty region, may discover that some phenomena just won't "fit the box." Two or more Good Ideas may turn out to be inconsistent, or various portions of the network may not fit together smoothly. Such problems spell trouble for normal science, and create the maladjustment that sets the stage for the second of Kuhn's kinds of scientific change—the *scientific revolution.*

By the 1870s, normal scientific work had pushed Newton's laws and the classical physics based on them so far that a professor of physics at the University of Munich warned the young physicist Max Planck that there was "nothing left to discover." The British physicist Lord Kelvin similarly suggested that "the future truths of Physical Science are to be looked for in the sixth place of the decimals." Nevertheless, within a few years theorists had come to the unsettling conclusion that the principles of classical physics imply that all objects should at all times be emitting an immense amount of ultraviolet light—so much that you would burn your eyes out just by opening them in a darkened room. This absurdity became known as the "ultraviolet catastrophe." Most scientists supposed that eventually, in the hands of some clever researcher, this puzzle would evaporate, and classical ideas would be vindicated. But after several decades of repeated failure, things began to look rather more ominous, especially as a host of similarly stubborn problems brought the maladjustment to the breaking point.

If a paradigm normally gives a scientist his or her foundations, its undermining leads to predictable distress. In the 1920s, the physicist Wolfgang Pauli, troubled by the tangles in the classical paradigm, wrote:

> At the moment physics is again terribly confused. In any case, it is too difficult for me, and I wish I had been a movie comedian or something of the sort and had never heard of physics.[17]

But when normal science runs aground, this also provides scientists with an opportunity. Normal science is conservative and considers paradigmatic ideas as virtually unalterable. Consequently, it is only the stress of considerable maladjustment that brings scientists to the point where they will consider digging up or rejecting some of these Good Ideas, and rebuilding their foundations anew. As Kuhn pointed out, it is a general pattern in science:

> . . . normal science repeatedly goes astray. And when it does—when, that is, the profession can no longer evade anomalies that subvert the existing tradition of scientific practice—then begin the extraordinary investigations that lead the profession at last to a new set of commitments, a new basis for the practice of science. The extraordinary episodes in which that shift of professional commitments occurs are . . . scientific revolutions. They are the tradition-shattering complements to the tradition-bound activity of normal science.[18]

In the case of physics in the 1920s, it fell to Werner Heisenberg, Erwin Schrödinger, and Paul Dirac, stimulated by the earlier ideas of Einstein, Planck, Niels Bohr, and Louis de Broglie, to tear up the intellectual landscape and lay the new foundations of quantum theory. After an extraordinary episode of this sort, scientists again have a firm paradigm to work from, and normal science can resume, even if the network has been radically altered. Compare Pauli's alarm with his restored confidence some months later, after Heisenberg had achieved his first tentative results at forging a new quantum paradigm:

> Heisenberg's type of mechanics has again given me hope and joy in life. To be sure it does not supply the solution to the riddle, but I believe it is again possible to march forward.[19]

The emotive word is "marching." For normal science is rather like marching, surefooted and confident, over familiar terrain.

To summarize Kuhn's perspective, normal science fills in and extends the existing network of Good Ideas, and does not aim in any way to produce any fundamental revisions in how scientists see the world. Ironically, however, this normal work itself inevitably turns up anomalies and inconsistencies, and leads to the growth of an internal stress within the existing fabric of ideas. And when this maladjustment reaches some threshold, that fabric, and the normal science based on it, breaks down. Scientists then find that they cannot go further by accumulation and extension, but have to tear apart and rebuild some portion of the existing network.

Reconstruction like this can never be completely isolated, of course. As in the Earth's crust, where the slipping of a few rocks alters the stress on nearby rocks and can trigger a traveling wave of further activity, so too does the rebuilding of one portion of the network necessitate further changes in neighboring regions. And these changes may, in turn, require still further changes elsewhere. The creation of the quantum theory of the atom, for example, meant that the scientific theories for solids, liquids, and gases had to be similarly rebuilt.

The Physics of Revolution

Kuhn's picture of science has been immensely influential. Of *The Structure of Scientific Revolutions*, the historian Peter Novick has written:

> It would be hard to nominate another twentieth-century American academic work which has been as widely influential; among historical books it would appear to be without serious rival.[20]

This is perhaps due to the fact that Kuhn's work did not result in mere narrative, but in a generalization that appears to apply to all cases of scientific change. In the tension between the "tradition-bound" and "tradition-shattering" modes of change, he identified the crucial elements of a deeper historical process. But Kuhn did not know the mathematical physics that might have enabled him to recognize just how deep and universal this process may be. If the basic elements of the Kuhnian pattern seem familiar, it is no surprise—they are strikingly similar to those lying behind the dynamics of earthquakes.

In the Earth, the slow movement of the continental plates does not directly bring about any reorganization of the Earth's crust, since friction holds the rocks in place. The movement of the continental plates simply puts the rocks under stress. Only when the stress builds up past some threshold do the rocks move and reorganize themselves, suddenly and violently. Similarly, normal science builds up stress in the network of Good Ideas. As Polanyi pointed out, the community of scientists possesses a kind of "mental friction," and the system of scientific ideas shifts in a revolution only when the stress passes a threshold.

Normal scientific work, then, is analogous to the drifting of the continental plates, and scientific revolutions are akin to earthquakes. And the analogy can be carried still further. As we know, earthquakes have no typical size. When those first few rocks slip, they may alter the forces on others nearby and so cause further slipping; and because the Earth's crust is naturally organized in a critical state, how far each such chain of slipping events carries on is entirely unpredictable. There is no typical earthquake. Could this also be the case with scientific revolutions?

In the names Albert Einstein, Isaac Newton, Charles Darwin, and Werner Heisenberg we have labels that identify the great, earth-shattering scientific revolutions. But in a 1969 postscript to a new edition of *The Structure of Scientific Revolutions*, Kuhn emphasized that such revolutions need not have widespread consequences or involve fundamental ideas. A small subfield of physics, say, or even a handful of scientists in a single research group, can experience a revolutionary change in the structure of ideas that form the basis of its work. If the ideas on which it has depended increasingly fail to deliver results, even a small group may experience the same basic pattern of change:

> Partly because of the examples I have chosen and partly because of my vagueness about the nature and size of the relevant communities, a few readers of this book have concluded that my concern is primarily or exclusively with major revolutions such as those associated with Copernicus, Newton, Darwin, or Einstein. . . . A revolution is for me a special sort of change involving a certain sort of reconstruction of group commitments. But it need not be a large change, nor need it seem revolutionary to those outside a single community, consisting of perhaps fewer than twenty-five people. It is just because this type of change, little recognized or discussed in the literature of the philosophy of science, occurs so regularly on this smaller scale that revolutionary, as opposed to cumulative, change so badly needs to be understood.

So in Kuhn's own mind, what distinguishes a revolution from normal science is its *tradition-shattering* as opposed to *tradition-preserving* character. A revolution tears up part of the old network of ideas; normal science simply adds to it. Kuhn's discussion hints at the possibility that there may be no typical revolution, that the dynamics of science could be scale-invariant, and that the network of Good Ideas might, like the Earth's crust, be poised in a critical state. But these are merely *possibilities*. An analogy does not prove anything. Is there any way to find more telling evidence?

· 11 ·

A Question of Numbers

Science . . . cannot exist on the basis of a
treaty of strict non-aggression with the rest of society;
from either side, there is no defensible frontier.

—JOHN KRASHER PRICE[1]

———————

There is nothing like a revolution to create
an interest in history.

—EDWARD HALLETT CARR[2]

✦

IT MAY BE IMPOSSIBLE TO GET ANY FIRM MATHEMATICAL HOLD ON something as elusive as the network of ideas at the core of science. It is one thing to strap the hills of California with delicate sensors to monitor the sticking and slipping of continental plates along the San Andreas Fault. But the rocky plates of the Earth's crust are out there, waiting to be measured; by contrast, the network of scientific ideas lives in the far less accessible realm of scientists' own thoughts and memories.

Even so, at one point in *The Structure of Scientific Revolutions*, Kuhn makes the following intriguing suggestion:

> . . . if I am right that each scientific revolution alters the historical perspective of the community that experiences it, then that change of perspective should affect the structure of post-revolutionary textbooks and research publications. One such effect—a shift in the distribution of technical literature cited in the footnotes to research reports—ought to be studied as a possible index to the occurrence of revolutions.[3]

Kuhn took the idea no further, but it is not too hard to see how it might be done. Any scientist lives and breathes the ideas of a certain area of science—particle physics, genetics, cosmology, or what have you—and, when publishing a paper, offers a list of citations as a way of locating his or her new ideas within the local network of Good Ideas for that specialty. In some more or less indirect way, the citations link the research papers in this area together in a way that reflects the structure of the network of ideas, even if the ideas themselves inhabit the ethereal atmosphere of the human mind.

[197]

We want to examine the nature of changes in this network, and, fortunately, citations also offer a way of doing this. We can take a clue from the geophysicists. In an earthquake, Earth scientists record the strength of ground-shaking as a measure of the earthquake's magnitude. This magnitude, in turn, reflects how much physical re-arrangement took place in the fabric of rock in the Earth's crust: larger earthquakes rearrange the landscape more extensively than do smaller ones. As we saw earlier, it was by studying the statistics of the magnitudes of many earthquakes that Gutenberg and Richter found their remarkably simple power law, a law that we now understand as implying that the origins of all quakes are essentially the same. The action always starts when the rocks along a tiny segment of a fault begin to slip. How big the quake turns out to be depends not on the triggering event, but on where it happens; on whether it trips only a short chain of slipping events in the rocks farther afield, or taps into one of the long "fingers of instability" that runs a long way through the crust.

By analogy, each scientific research paper is a package of ideas that, when it nestles down in the preexisting network of ideas, triggers some larger or smaller rearrangement. It may be a theoretical paper, something like Watts and Strogatz's paper on small worlds. This paper identified an unexpected connection between the mathematics of graphs and the peculiar properties of social networks. This new idea has by now altered some of the beliefs and research interests of other scientists. Some have written papers exploring in more detail the mathematical properties of small-world graphs; others have begun applying the basic mathematical insight to understanding the spreading of diseases, and so on.

It is too early to tell how much further activity the ideas in this one paper will ultimately stir up. But it is likely that many of the papers that result from such activity will cite the original small-worlds paper. So to measure the overall size of the intellectual earthquake triggered by this or any other paper, we might look to the total number of times it is cited by other, later papers. A paper that garners only one citation causes very little reorganization in the network of scientific ideas; one that receives a thousand causes a lot.

This is of course only a crude way to measure the ultimate effect of a paper. Still, we might ask a question à la Gutenberg and Richter: What is the typical number of citations received by a paper? That is, when a new idea comes along, what is the typical size of the intellectual earthquake it triggers?

The Paper Trail

Fortunately, it is quite easy to track the citation history of a paper. The Science Citation Index is a resource listing all the citations ever received for any scientific research paper dating back into the 1960s. Pick out some random paper in quantum field theory published in December 1967, and you can find out who from that date onward has cited the paper. In 1998, physicist Sidney Redner of Boston University did this, not just for one paper, but for all 783,339 papers published in 1981. It is necessary, of course, to consider papers published some years ago, so that those destined to collect a mass of citations will have had time to do so. Otherwise, the numbers won't truly reflect the response each paper engendered.

Looking into the statistics for his selection of papers, Redner first discovered a rather sobering truth: a full 368,110 of these were simply never cited. The ideas in these papers caused no perceptible response whatsoever in the network of ideas. Looking at papers that had more effect, however, Redner found something more interesting. For papers that collected more than a hundred citations or so, the distribution of citations follows a scale-invariant power law, precisely what we might expect if the network of ideas is organized, like the sandpile game or the Earth's crust, into a critical state. Papers that receive a higher number of citations are, as expected, less numerous than those garnering fewer. But Redner discovered that

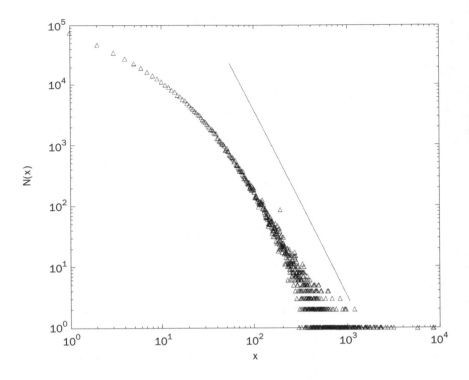

FIGURE 19. Distribution of research papers according to the number of citations they receive. Adapted from Sidney Redner, *How Popular Is Your paper? European Physical Journal B* (1998) 4: 131–134, reprinted by permission.

the way highly cited papers become rare follows an extremely regular pattern: double the number of citations, and the number of papers receiving that many falls off by about eight (SEE FIGURE 19). So there is no typical number of citations for a paper, and, by extension, no typical magnitude for the reshaping in the network of ideas that any paper ultimately entails.[4] What does this imply?

Earlier, in considering mass extinctions, we saw that the tremendous extinctions stood out against the background of the more run-of-the-mill extinctions. To the naïve eye, the two kinds appeared to arise from essentially different causes: external shocks, perhaps, as opposed to the ordinary workings of evolution. It turned out, however, that this distinction was illusory. We saw the same thing in the case of earthquakes: however strongly we are drawn to look for exceptional causes behind the great quakes, the Gutenberg-Richter law implies that no such exceptional causes exist. In the light of Redner's statistics, it looks as if something very similar may be true of science itself. When Kuhn hinted that there could be small revolutions as well as large, and that both shared the same essential character as "tradition-shattering" events, this was as far as he could go. But Redner's power law for citations is a kind of Gutenberg-Richter law for scientific upheavals, and implies that there is, in a deep sense, no true distinction between large and small scientific revolutions.

Together with this "intellectual" Gutenberg-Richter law, Kuhn's landmark analysis suggests that the fabric of scientific knowledge, like the Earth's crust and so many other things, is poised in a critical state. If so, then scientists should expect the unexpected. For the fabric of ideas is organized so that the tiniest chance discovery might at any moment and without warning activate a domino-like chain of effects leading to a terrific revolution. And foreseeing such revolutions is next to impossible, for the ultimate consequences of any new idea depend not so much on its own inherent profundity as on where it happens to fall within the network of all scientific ideas.

Everyone associates Einstein's name with one of the greatest revolutions in scientific history. But the Einsteinian revolution started when Einstein began puzzling over a quirky feature of Maxwell's equations, which describe light as an electromagnetic vibration. These simple equations, he found, would not permit him to imagine riding along with a light wave and studying it as if it were "standing still." It was this minor conceptual paradox, almost a mere curiosity, that ultimately led to the revision of several hundred years of physics and the theory of relativity, and then, through myriad other pathways, to both nuclear energy and the atomic bomb.

Similar examples can easily be multiplied. In 1900, Planck discovered a formula to account for the colors given off by a hot, glowing object, and to arrive at the formula he had to suppose that when light interacts with matter, energy can only go from one to the other in tiny, discrete chunks. At the time, Planck thought this was merely a cheap, technical trick that gave the right answer, probably for the wrong reason. Neither he nor anyone else had any inkling that this technical trick would turn out to be toxic to nearly all of physical science, and would lead ultimately to the staggering revolution of quantum physics.

As seen through the lens of the critical state, great revolutions are not necessarily unique in terms of their causes. They are simply the expected "large fluctuations" of a system poised in a critical state.

The Sandpile of Science

This is not to suggest that one scientist is not more capable than another, that Einstein was an average guy, or that his 1905 paper on relativity was not especially profound. I have been speaking so far as if the avalanches that run through the fabric of ideas always do so by moving from one mind to another. But such avalanches can also take place in single minds, when an individual comes upon one idea that he or she realizes has consequences for other ideas, and is able to draw those consequences clearly, possibly discovering along the way still further consequences for other ideas. Lars Onsager experienced this in his work on the two-dimensional toy magnet:

> It was the sort of investigation where you got a good lead, and certainly you had to pursue that; and before you reached the end . . . up opened another . . . one lead opened up after the other, every one much too good to abandon.[5]

Similarly, the Einsteinian revolution started out in Einstein's own head with the peculiarity of Maxwell's equations, but before the avalanche had ever left Einstein's own head it had already moved a very long way. Perhaps no scientist except Einstein would have gone so far on his own. One might well suppose, then, that the mark of the great scientist lies not so much in having profound ideas that revolutionize science, but in taking ideas that have the potential to do so and in making that potential real. In the Earth, the rocks slide on their own, pushed by the enormous stress built up in the crust. In the sandpile, gravity makes grains tumble. But scientific changes come about only through scientists' hard work. The great scientist has skill in locating those ideas in the fabric that have the potential for setting off domino-like chains that will at least extend an appreciable way, and the ability and energy to make those repercussions explicit.

So where does this leave us? In the nineteenth century, people could speak of science as a process akin to the building of a great tower of knowledge. Scientists were an army of builders, each contributing small bricks and placing them into their proper positions in the growing tower. Kuhn's historical and philosophical work showed this was a drastic oversimplification. Science does not grow by simple aggregation. Sometimes when a new brick gets put in place, it points out such a flaw in the architecture of what already exists that scientists need to tear down a portion and rebuild it before they can begin again.

In the light of Redner's power law, we can make this image a bit more definite. Every new idea of science that pops into a theorist's head, or every observation made by an experimenter, is something like a grain falling on the pile of knowledge. It may stick, and merely add to a growing structure, or it may place a portion of the pile under such stress that ideas will topple. The toppling may stop quickly or may run on for a long while. As reflected in the scale-free power law for citations, the avalanches have no inherent or expected size. The smallest revolutions are happening every day, may involve only a few scientists in specialized communities, and may be virtually invisible to almost everyone else, just like those tiny earthquakes going on all the time beneath our feet. By contrast, the largest revolutions may wipe away much of science as we know it, and are liable to happen at any moment, if the right idea pops up in the right place.

At the beginning of the last chapter, we looked at the shortcomings of the narrative approach to understanding anything in which history matters. Wherever contingency rules, any tiny accident can shunt the future irrevocably down one route or another, and so when it comes to explaining the course of complex chains of events, there can be no simple, deterministic laws. This is true in human history, and also in the sandpile game and in the Earth's crust. And yet in the latter cases, there is clearly more to be said. We know that these things share the character of the critical state, reflected in remarkably simple, statistical laws: the scale-free power laws that reveal a profound hypersensitivity built into the system and the lack even of any expected size for the next event. So while chains of events in these systems may not be predictable, it is not the case that nothing is predictable. It is in the statistical pattern that emerges over many chains of events that we can hope to discover the laws for things historical.

Such laws capture the general properties of many narratives, rather than just one, and thus reflect the character of the deeper historical process that operates behind individual chains of events. And remarkably, such a law also lies behind the historical dynamics of science. In the way scientists work, in the way their ideas interact and lead to new ideas, there is a natural organization that enables tiny initial causes to be amplified enormously. The law for intellectual earthquakes reflects how easy it is for influences to "propagate," even if these influences are, as in this case, as ethereal as pure ideas.

Ideas affect a lot more than just science. Every area of human activity from city planning to the theater has its own ecosystem of interacting Good Ideas (which would include practices, techniques, and so on). So too in art, fashion, or music. And when a new idea comes along, it doesn't settle into the fabric of preexisting ideas without repercussions. So by analogy we might expect ideas in almost every field to evolve according to similar statistical laws and with the same kind of sporadic rhythm as they do in science. There ought always to be times when everything seems settled, and when it seems that everything possible has already been done. But occasionally, a minor twist on some old idea—nothing too dramatic at the outset—ought to trigger a spreading wave of rethinking that ends up as a world-shaking revolution.

More boldly, we may even expect to find the power-law signature of the critical state in the mighty river of human history considered more broadly.

Human, All Too Human

British Prime Minister Winston Churchill was once asked what the desirable qualifications were for any young person who might wish to become a politician. He replied:

> It is the ability to foretell what is going to happen tomorrow, next week, next month, and next year. And to have the ability afterwards to explain why it didn't happen.[6]

History indeed is unpredictable. On this, nearly all politicians and historians agree. Nevertheless, most also share the conviction that the course of human affairs does not suddenly turn around or go mad for no good reason. When war breaks out, or when the chaos of revolution or economic catastrophe engulfs a nation, historians proceed in the faith that the causes of such events can be identified, like those of a disease in the body.

It is not comforting to consider the alternative: that the world can explode for no good reason, and that even while the skies look clear, largely invisible forces may nevertheless be conspiring to undermine the fabric of a society, or of the relations between nations, and visit calamity upon us in the near future. At the outset of this book, I suggested that theoretical physics is beginning to offer hints as to why history has the unruly character that it does, and, especially, why it is and perhaps even has to be shattered irregularly but frequently by momentous and unpredictable upheavals. Anyone who has read this far can probably begin to understand why this might be the case. To take one final step toward human history more generally, it will help to make a few final comments about Kuhn's image of science, and his insight into the nature of human thinking that lies behind it.

Kuhn's principal achievement was to show that science works even though scientists, like everyone else, labor under the burden of being human. Rather than being a superrational machine, each scientist struggles with his or her share of blind ambition, prejudice, bias, and timidity, all fueled by that craving for certainty that is distinctively human. In every paradigm, scientists discover a logical structure that makes sense of a portion of the world and gives them an intellectual foundation to which they cling until the discrepancies and inconsistencies grow so unsettling as to force them to break with tradition and alter some of their cherished ideas.

This is not quite enough to give rise to a critical state and the power law for scientific revolutions. But Redner's power law teaches us something else about the nature of scientific change. In the sand-pile game, when a portion of the pile becomes too steep, sand slides downhill, but only until the slope again settles just slightly below the threshold. That is, while an avalanche reduces the stress, it also maintains the pile in the critical state and on the edge of instability; and this, ultimately, is what lies behind the lack of any typical scale for such avalanches. The same is true in the Earth's crust, where the slipping goes on until the friction in the rocks somewhere is just barely sufficient to bring things to a halt.

Redner's power law for citations tells us that something similar is true, at least in some crude sense, of the dynamics of scientists' thinking. Scientists driven to consider altering a few crucial bricks in their theoretical foundation do not recklessly demolish the entire house, but remake their foundation only insofar as is absolutely necessary. As in the sandpile game, this naturally maintains the fabric of belief so that the stress everywhere is approaching the rupture point, in which case the next tiny crisis just might, by domino-like action, lead to a revolution out of all proportion to itself. In a nutshell, it is intellectual friction that holds the network of ideas in place, and intellectual curiosity that puts it under stress. These two influences play against one another as do friction and the relentless driving pressure of continental drift in the case of earthquakes. And it is this competition that leads to the critical state and the power law for scientific revolutions.

But scientists, of course, are not alone in craving certainty, or in being loath to begin shattering their traditions. This is the way in which scientists behave as human beings, and this way of behaving is archetypal for all humanity, under all conditions. These considerations hint at the possibility that Kuhn was tracing the silhouette of a pattern of universal change that runs far deeper than he suspected. And indeed it is not hard, at the level of historical description, to find this logical skeleton elsewhere.

Civilization and Its Discontents

As I mentioned in chapter 1, the historian Paul Kennedy has suggested that the large-scale historical rhythm in the interactions among the Great Powers is largely a consequence of the natural buildup and release of stress driven by national interests. According to Kennedy:

> The relative strengths of the leading nations in world
> affairs never remain constant, principally because of the
> uneven rate of growth among different societies and of

the technological and organizational breakthroughs which bring a greater advantage to one society than to another. For example, the coming of the long-range gunner sailing ship and the rise of the Atlantic trades after 1500 was not uniformly beneficial to all the states of Europe—it boosted some much more than others. In the same way, the later development of steam power and of the coal and metal resources upon which it relied massively increased the relative power of certain nations, and thereby decreased the relative power of others.[7]

Such natural changes leave some nations clinging to power that their economic base can no longer support; others find new economic strength, and so seek greater influence. Tension inevitably grows until it passes some threshold—perhaps as the result of some momentary and largely accidental crisis—and something gives way. Usually, the stress finds its release through armed conflict, after which the influence of each nation is brought back into rough balance with its true economic strength.

A roughly similar pattern probably underlies the interactions among the various groups and individuals within any nation. No society remains fixed, and as it changes some groups, even if by historical accident, become more powerful than others, triggering the development of internal problems—economic, racial, or what have you. Any society has tradition-bound structures, such as social conventions, moral prohibitions, class structures, and laws, aimed at maintaining stability and negotiating the conflicts among its members. But the tradition-bound structures are not always adequate.

Just as science is not all normal science, politics is not all government as usual. Kuhn himself pointed to the analogy:

> Political revolutions are inaugurated by the growing sense . . . that existing institutions have ceased adequately

to meet the problems posed by an environment that they have in part created. In much the same way, scientific revolutions are inaugurated by a growing sense . . . that an existing paradigm has ceased to function adequately in the exploration of an aspect of nature to which that paradigm itself had previously led the way. In both political and scientific development, the sense of malfunction that can lead to crisis is prerequisite to revolution.[8]

Even so, because of the human craving for stability, especially on the part of those who are in power and who benefit most from the existing order, nothing will give way in the fabric of existing institutions until the strife and discontent build beyond some threshold. The people do not rise up in revolution until their discontent becomes so great that they have no other recourse. People do not take up active protest against what they see as unjust laws unless those laws cause them considerable discomfort. Recall again the words of the American historian Conyers Read:

> Unless we are alert to the necessity of constant re-adjustment we create a condition of maladjustment which is the inevitable forerunner of Revolution, whether that Revolution take the Russian or the Italian form. . . . I believe that the study of history has an important social function to perform of just this sort.

We might well compare Read's lesson to that learned by forest managers in the American West, who are now similarly alert to the benefits of "constant re-adjustment" as it takes place naturally in the forests through smaller fires. Preventing such "adjustments" merely makes matters worse.

None of this is meant to be fully convincing. Neither the "web of international relations" nor the social fabric of any society is a thing easily grasped. Even so, it is probably beyond argument that some form of stress must build up in both of these things. It seems likely as well that as this stress builds, it is not always or even often immediately released by "adjustments." It has to build beyond some threshold before anything changes. Tradition within any community is a powerful form of social friction. So stress builds and friction holds things in place, until finally giving way.

As we have seen many times already, it is quite natural under such conditions that interactions among all the various elements making up any system tend to organize it into a state of criticality. In this condition, a sudden release of stress in one place can trigger an avalanche of further releases that can travel a long way. Proving that this is really the case in any social system may be very difficult. If it is true, however, then we should expect that the world will necessarily—and not infrequently—suffer major wars, and that any society will experience tumultuous revolutions, events that will seem to emerge almost from nowhere.

Intriguingly, there is at least a tiny bit of mathematical evidence to suggest that the world really is arranged this way.

A Grim Reckoning

From his study of the past, Fyodor Dostoyevsky drew a simple lesson:

> They fight and fight and fight; they are fighting now, they fought before, and they'll fight in the future. . . . So you see, you can say anything about world history. . . . Except one thing, that is. It cannot be said that world history is reasonable.[9]

Out of all that fighting, the most terrible wars stand out head and shoulders above the rest. Like the great earthquakes or the great mass extinctions, they seem special and exceptional. But are they really? That is, were the great wars set up and set off by special and unusual conditions that men with greater foresight might have recognized?

In science, the record of citations closely tracks alterations in the fabric of ideas, and makes it at least possible to put some crude numbers on the process. It is more difficult in the case of general history. In the 1920s, however, the British physicist Leslie Richardson studied eighty-two wars that had flared up between 1820 and 1929. Of all the possible ways of judging the size of a war, Richardson chose the most obvious and most grim: the number of deaths. Following a procedure like that of Gutenberg and Richter, he counted up how many wars caused between five thousand and ten thousand deaths, how many between ten thousand and fifteen thousand, and so on, and then made a graph. He obtained a curve showing how often wars of various sizes took place, and found a very simple power law: every time he doubled the number of deaths, he found that wars of that size became 4 times less common (SEE FIGURE 20). This is precisely the same as the Gutenberg-Richter law, and we can draw similar conclusions. There is no typical size for a war, and no sensible division of wars into small skirmishes and major conflagrations. All wars fall on a smooth curve that implies that the initial causes of all wars are probably the same.

The human population is today, of course, far larger than it was several centuries ago or even one century ago. So one might raise an obvious objection to Richardson's style of analysis: Might the power law not simply reflect the fact that in recent times conflicts have become far more deadly, there being more people available to become involved in them? This is a legitimate objection, and yet other researchers have found a similar power law even when correcting for changes in population. In the 1980s, for instance, Jack Levy of the University of Kentucky looked at wars starting with the War of the League of Venice in 1495, and ending with the Vietnam

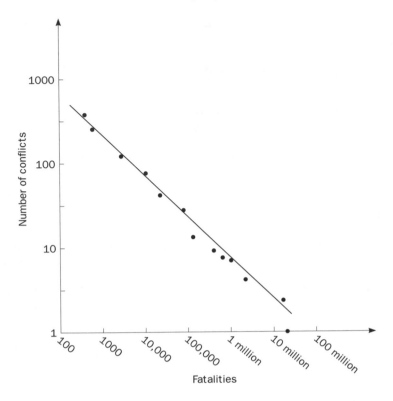

FIGURE 20. Distribution of deadly conflicts according to the number of people killed.

War in 1975, and altered Richardson's prescription by taking the size of a war to be the number of casualties divided by the population at the time. In other words, he took the size of a conflict as the fraction of the population killed. Even so, he again found a power law,[10] although the precise numerical pattern is very slightly different: viewed in this way, wars become about 2.62 times less frequent every time you double the number of deaths.

It is a truism to say that all wars are triggered by conflict. But the power law implies that a war, when it starts out, does not "know how big it will become." Nor does anyone else. Somewhere, for some reason, the intensity of disagreement, competition, distrust, or hatred between one group and another passes a threshold, and they prepare to take up arms. The traditional structures that negotiate the differences between men rupture, and they resort instead to brute physical force. Whether that rupturing spreads and the war becomes far larger depends on whether neighboring regions also happen to be near the rupturing point, and make it possible for the trouble to carry over into other minds, other communities, other nations. This in itself is hardly news; every analyst of international events is concerned that trouble in one region may "destabilize" neighboring regions, and the international responses of the United Nations, the North Atlantic Treaty Organization, and other such organizations are largely aimed at heading off such destabilization.

What is news is the power law, which suggests that the world's political and social fabric tends to be organized on the very edge of instability, and in such a way that wars spread in a special way, so that their ultimate extent is nearly impossible to foresee. The scale-free character of the power law suggests that at the outset of a war there are no obvious clues about how large it will become. The organization of the ties that bind us peacefully into nations and communities appears to be such that war spreads like a forest fire, or like an avalanche in the sandpile game.

Indeed, geophysicist Donald Turcotte of Cornell University has pointed out that the number 2.62 that appears in Levy's power law for European conflicts is startlingly close to the numbers that tend to come out of the forest fire game, which tend to be between 2.5 and 2.8. So it is tempting to conclude that the forest fire game captures the crucial elements of the way that conflicts spread. As Turcotte has speculated,

> . . . a war must begin in a manner similar to the ignition
> of a forest. One country may invade another country, or
> a prominent politician may be assassinated. The war may
> then spread over the contiguous region of metastable

countries. Such regions of metastability may be the countries of the Middle East (Iran, Iraq, Syria, Israel, Egypt,
etc.) or of the former Yugoslavia (Serbia, Bosnia, Croatia,
etc.). . . . Some fires are large and some are small. But
the frequency-size distribution is power-law. In terms of
world order there are small conflicts that may or may not
grow into major wars. The stabilizing and de-stabilizing
influences are clearly very complex.[11]

In view of this remarkable correspondence between the patterns by
which wars and forest fires spread, it is no surprise that historians
have been unable to agree on the cause of a conflict such as the First
World War. Those who pointed vaguely to a "breakdown of the
international system" may at least have been on the right track.
Perhaps the war's cause was not to be found in any specific trigger,
but lay instead in the overall organization of the fabric of human
relations, sociological, economic, and political, which made it possible for malicious influences "to propagate."

Indeed, although he may not yet appreciate how right he may be,
one perceptive historian has recently gone so far as to suggest that
the First and Second World Wars were, in essence, like a pair of
earthquakes:

> . . . the years between 1914 and 1945 appear as the time
> of Europe's troubles, which filled the space between the
> long peace of the late nineteenth century and the still
> longer peace of the "Cold War." They may be likened to
> the slipping of a continental plate, and to the resultant
> season of earthquakes. They encompass the initial mili-
> tary quakes of 1914–1918, the collapse of four empires,
> the outbreak of communist revolution in Russia, the
> emergence of a dozen new sovereign states, the armed
> truce of the interwar decades, the fascist take-overs in

Italy, Germany and Spain, and then the second, general military conflagration of 1939–45.[12]

The Power of Persuasion

When it comes to other important events, such as revolutions, it is rather more difficult to find any suggestive mathematical evidence. While the Chinese, French, and Russian revolutions were murderous affairs, equally profound changes have swept over both South Africa and the former Soviet Union with very little violence. So no appeal to the statistics of casualties is likely to be very informative. Moreover, revolutions needn't even be political. There are revolutions in art and music, in social conventions and working habits, technological revolutions, and so on. But behind all important social changes there ultimately lies one simple driving force: the ability of one human being to influence another. As we saw earlier in the case of the financial markets, this feature, while entirely obvious, can nevertheless be profound in its effects.

It is not denying free will to observe that people very commonly buy specific products, adopt certain opinions, vote one way or another, or take to the streets in protest largely because their friends, neighbors, family, or coworkers have done the same. In December 1989, hundreds of thousands of Romanians took to the streets in protest again the totalitarian regime of Nikolae Ceauçescu. It is inconceivable that they all simultaneously but independently made up their minds to do it. All mass movements start out small and then swell, as the actions of a few infect others and the movement spreads like a virus. In the financial markets, mass movements lead to dramatic changes in the prices of stocks or bonds. In politics, they sweep governments into or out of office, precipitate revolutions, or take nations to war.

This is all obvious, and historians have talked about it for centuries. Just as it is not news that strife in one country can spill over into another, it is similarly an old story to historians that great social movements are triggered not so much by one individual as by the collective effects that take place among many. In the words of Carr:

> The facts of history are indeed facts about individuals, but not about actions of individuals performed in isolation, and not about the motives, real or imaginary, from which individuals suppose themselves to have acted. They are facts about the relations of individuals to one another in society and about the social forces which produce from the actions of individuals results often at variance with, and sometimes opposite to, the results which they themselves intended.[13]

For Carr, collective dynamics were central to history. Nevertheless, when he was writing, he had no way of knowing that these dynamics might, at least in principle, arise from the critical state, and so be unruly in a profound way.

From this perspective, it becomes far easier to understand how mass movements take off. To understand any particular revolution, historians surely need to study all the social conditions from which it springs. To understand what makes a person take up arms, go on strike, or decide not to have children, the historian indeed has to try to get inside that person's mind, and weigh up all the social pressures and influences to which he or she is responding. Only in this way can the historian come to understand what sparked a revolution, as many people's actions followed in some understandable way from the conditions they were in. But the historian really needs also to know more about the way in which influences of all kinds can propagate through a population. To understand why mass movements are not rare, and history is as interesting and as varied as it is, we need to understand the character of the critical state.

In the 1920s, Bohr got a bit carried away with the uncertainty principle of quantum theory, which proclaims that the very act of observing a quantum particle inevitably disturbs its properties. He even went so far as to write papers suggesting that this principle might usefully be extended to apply in the social sciences, or in psychology, where the observer similarly has an unavoidable effect on the behavior of the observed. Well before Bohr, other thinkers had tried to import ideas from physics into the humanities by appealing to their own mangled misinterpretations of Einstein's theory of relativity. Einstein himself thought there was something "psychopathological" about such efforts:

> I believe that the present fashion of applying the axioms
> of physical science to human life is not only a mistake but
> has something reprehensible to it.[14]

But both quantum theory and relativity are based on timeless equations, and in no way involve history in any form. In striking contrast, in the sandpile game and the other simple games that we have met, history matters in a crucial way. And if the character of the critical state tells us something deep about the way in which influences can propagate and carry order, disorder, or change through any collection of interacting things, it is not ludicrous to suppose that sociologists and historians might even find it a valuable concept. If we don't get carried away, and don't try to draw too many lessons, the critical state may indeed be able to give us a few hints about how we should expect human history to unfold.

· 12 ·

History Matters

It is always a mistake for the historian
to try to predict the future. Life, unlike science,
is simply too full of surprises.

—RICHARD EVANS[1]

———————

There are no foolish questions and no man
becomes a fool until he has stopped asking questions.

—CHARLES PROTEUS STEINMETZ[2]

IT WAS THE EMINENT BRITISH HISTORIAN THOMAS CARLYLE WHO first declared, "The history of the world is but the biography of great men." To historians who think this way, it was Adolf Hitler who caused the Second World War, Mikhail Gorbachev's genius that by itself brought the Cold War to an end, and Mahatma Gandhi who won independence for India. This is the "great person" theory of history—the view that sees exceptional human beings as standing virtually outside the stream of history, and imposing their will upon it "in virtue of their greatness."[3]

This way of interpreting history certainly has an appeal, in part, no doubt, because it makes the past seem fairly simple. If Hitler's evil was the ultimate cause of the Second World War, then we know why it happened and whom to blame. We also know how to avoid such trouble in the future. If someone had strangled Hitler in his crib as an infant, there would have been no war, and millions of people would have been spared. History, from this perspective, is simpler than it might otherwise be, for the historian need only follow the actions of a few great actors, and can safely ignore everyone else.

Many historians have been of another mind, and have seen in this view little more than a grotesque parody of how history really works. "Nothing causes more error or unfairness in man's view of history," wrote Lord Acton in 1863, "than the interest which is inspired by individual characters."[4] Edward Hallett Carr similarly repudiated the "great person" theory of history as being "childish" and characteristic of "the primitive stages of historical thinking":

> It is easier to call communism the "brain-child" of Karl
> Marx . . . than to analyze its origin and character, to
> attribute the Bolshevik Revolution to the stupidity of

Nicholas II or to German gold than to study its profound social causes, and to see in the two world wars of this century the result of the individual wickedness of Wilhelm II and Hitler rather than of some deep-seated breakdown in the system of international relations.[5]

For Carr, the really important forces in history were the social movements, initiated perhaps by individuals, but important only because they involve huge numbers. "History," he concluded, "is to a considerable extent a matter of numbers."[6]

Great men and great women do exist, of course, and they have decisively influenced the course of history. Every mass movement has its leaders. But for historians of this more collective persuasion, it is a mistake to think that these leaders orchestrate the unfolding of history. No individual lives or thinks in a vacuum; rather, every person is profoundly influenced by others. As a result, no individual is quite the self-possessed and independent actor that he or she may appear to be. As the French historian Alexis de Tocqueville put it:

> Among all civilized peoples the political sciences create, or at least give shape to, general ideas; and from these general ideas are formed the problems in the midst of which politicians must struggle, and also the laws which they imagine they create. The political sciences form a sort of intellectual atmosphere breathed by both governors and governed in society, and both unwittingly derive from it the principles of their action.[7]

So while great characters are at the center of great happenings, they do not supply the forces that drive them. Instead, the role is more important than the person who occupies it, for the role is the point where great social forces collide, and it is in filling such pivotal roles that great men or women become great.

Consider Hitler. If he had been strangled in his crib, or had otherwise died before coming to power, would there have been a war? Any "great person" theorist would say no. But any historian who regards collective forces as preeminent would see this as a far more complicated question, and might agree with the Cambridge historian Richard Evans, who recently suggested that the social and political setting in Germany was such that, had Hitler and the Nazis never come to power, the war would have happened anyway:

> The chances of the Weimar Republic surviving were very small after the Depression began in 1929, and a far-right dictatorship of someone like Franz von Papen . . . or a restoration of the Hohenzollern monarchy . . . would almost certainly have led to a similar sequence of events to that which took place anyway: rearmament, revision of the Treaty of Versailles, Anchluss in Austria, and the resumption, with more energy and determination than ever before, of the drive for conquest which had been so evident in Germany's war aims between 1914 and 1918.[8]

It would be silly to think that mathematical physics can settle the debate about whether "great people" do or do not decide the course of history. History is infinitely more complicated than any sandpile. And yet thinking about the sandpile can help to identify some errors that are all too easy to make when looking at history, and when trying to untangle its causal threads.

Great Grains

Suppose our sandpile historian, at the end of a long and distinguished career, takes on the task of writing a definitive text: *The History of the Sand World.* For inspiration, he or she might well look to the German historian Leopold von Ranke, who, in 1878, at the age of eighty-three, embarked on his own momentous *History of the World*, completing no less than seventeen volumes by the time he died eight years later. To complete such a grand history, our sandpile historian would want to consider the circumstances and effects of every single avalanche that had ever occurred. That being beyond him, he might focus more pragmatically on the largest of the avalanches in the past, leaving the smaller ones to other historians with narrower tastes.

This would be sensible. For the largest avalanches are far and away the most influential in terms of the effects they have on the pile. Indeed, the very largest may involve millions of grains, while the vast majority will involve only a few: numbers count in the sandpile. In this sense, the history of the pile has to be written in terms of the great social movements. But this does not answer the question, How should the historian explain these great movements?

Our historian will also be sorely tempted to identify certain individual grains as having been massively influential. After all, his colleagues will point out that in 1492, an individual grain of immense courage named Granular Columbus triggered an awesome avalanche that ultimately carried grains all the way from the East to the West, and so altered the entire face of their world and its future history. These same historians may blame some other character for having sparked off an infamous catastrophe that made half the grains in the East slide down the hill. For each great event, they can identify some standout grain that touched it off, and perhaps a few others that kept it going at crucial stages. And these grains, they might conclude, are the real agents of history.

Our historian might be tempted to agree, and yet, after such a long career, he may have learned a few things. As a subtle observer of individual character, he will have noticed that in his world every grain is identical to every other, so there really can be no question of any one being a "great grain." As a result, he realizes that whatever the psychological pull to identify great events with great grains, the idea is a mistake. Only by understanding that his pile is always on the edge of radical change can he resist this temptation. If he does his job well, our historian comes to understand that there are always places in the pile at which the falling of a single grain can trigger world-changing effects. These grains are only special, however, because they happened to fall in the right place at the right time. In a critical world, there are necessarily a few great roles, and some grains by necessity fall into them.

Might the same be true of human history? There is no denying that some people, by virtue of their personality or intelligence, are more influential than others. And yet it is at the very least a theoretical possibility that our world lives in something very much like a critical state. In such a world, even if human beings were identical in their abilities, a few would nevertheless find themselves in situations in which their ordinary actions would have truly staggering consequences. They may not even be aware of it, as the potential for their actions to propagate may become apparent only as history unfolds. Such individuals may come to be known as great men or great women, as creators of vast social movements of tremendous import. Many of them may indeed be exceptional. But this need not imply that their greatness accounts for that of the events they sparked off.

Just as it is almost irresistibly tempting to seek great causes behind the great earthquakes or the great mass extinctions, it is also tempting to see great persons behind the great events in history. But the sandpile historian comes down firmly against the "great grain" theory of history, and would counsel his colleagues in the human world to follow his lead. He might agree with Georg Wilhelm Friedrich Hegel, who concluded that

> the great man of the age is the one who can put into words the will of his age, tell his age what its will is, and accomplish it. What he does is the heart and essence of his age; he actualizes his age.[9]

In this view, the greatness of an event is not traceable to an equivalent greatness existing in some individual. Rather, what makes an individual notable and "great" is his or her ability to unleash pent-up forces—the will of an age—and so enable those immeasurably greater forces to have their effect.

In the context of science, Einstein was a genius of the first order. Because of his genius, he could draw the implications of Maxwell's equations before his contemporaries did. But the theory of relativity was revolutionary not because of Einstein's genius, but because it represented a terrific avalanche in the fabric of ideas. Even if scientists were all genetically identical clones, such revolutionary achievements would still be set off by a select few. To borrow a remark of the biologist Edward O. Wilson, "Genius is the summed production of the many with the names of the few attached for easy recall."

In human history more generally, it may similarly be that the ability of individuals to have great influence depends more on the special organization of social systems than on anything else. And as it turns out, this is probably true not only of the actions of individuals, but of many other quirky historical accidents as well.

Cleopatra's Nose

"Writing history," Gustave Flaubert once observed, "is like drinking an ocean and pissing a cupful."[10] Swirling around any historical event are so many facts that the historian, long before taking up his or her pen, faces the daunting task of selecting a very small fraction of historical facts as worth mentioning. There are facts to be told about the clothing that Hitler wore as a boy, or about how many times Margaret Thatcher has eaten fish and chips, but these aren't historical facts. Out of all the infinite sea of facts, only a very few inform the historian about the important events and undercurrents that drove history along its course.

But there is a problem. For as Henri Poincaré once pointed out, "the facts do not speak." They do not rise up out of that sea like whales and declare themselves to be noteworthy. To work out which facts are historical and which aren't, the historian has to impose himself or herself on them. If a historian sees political influences as being generally predominant over economic forces, then he or she will tend to fish the political facts out of the sea, leaving the others there. Another historian might fish differently. Carr suggested that all historians have bees in their bonnets, and that the reader should beware:

> When you read a work of history, always listen out for the buzzing. If you can detect none, either you are tone deaf or your historian is a dull dog. The facts are really not at all like fish on the fishmonger's slab. They are like fish swimming about in a vast and sometimes inaccessible ocean; and what the historian catches will depend, partly on chance, but mainly on what part of the ocean he chooses to fish in and what tackle he chooses to use— these two factors being, of course, determined by the kind of fish he wants to catch.[11]

Related to this problem of fact selection and its dependence on personal bias is another quandary that historians refer to as the "Cleopatra's nose" difficulty of history. Mark Antony was infatuated with Cleopatra and because of her beauty led his ships into battle and ultimately to defeat at the hands of Octavius at Actium. Any legitimate account of the origin of that battle, and of its consequences— including the founding of the Roman Empire—must refer to Cleopatra's beauty. As Pascal famously expressed it, "Cleopatra's nose, had it been shorter, the whole face of the world would have changed." Churchill once pointed to a version of the same peculiarity: In 1920, the King of Greece died after being bitten by a pet monkey. On the ensuing chain of events that led Greece and Turkey to war, Churchill commented, "A quarter of a million persons died of this monkey's bite." If the smallest of details continually intrude on the larger picture, with the power to alter it radically, how is the historian to make sense of anything? Faced with this dilemma, it becomes doubly difficult for the historian to separate the significant historical facts from the background of all facts.

Almost all historians have written about these issues at one time or another. Carr certainly did, and we ought to have a look at one of his proposals for how to make sense of things, as it is still being repeated and discussed by historians today. There is, Carr claimed, a natural hierarchy of facts. Consider a certain Mr. Robinson who crosses a road near a blind bend to get cigarettes and is killed by a drunk driver. What caused his death? Carr sought the noteworthy causes in those that can be generalized. If Robinson hadn't wanted cigarettes, he wouldn't have been killed. This is true. So his desire for cigarettes was a cause of the incident. It wasn't a general cause, however, since, in general, the desire for cigarettes doesn't tend to lead to people being run over. By contrast, drunk driving and blind bends—other contributing causes of Robinson's death—can be generalized. Drunk driving and blind bends do lend themselves to people being run over, and so ought to be considered as the significant causes of the incident. Similarly, noses or monkey bites aren't general causes of nations going to war, although they may be contributing causes. For Carr, history was mainly about the general causes, as these offer lessons.

In this way, Carr segregated the causes of historical events into a secondary division of curious accidents and a primary division of general causes, the material of what history is really about. All this is quite reasonable. But how does it work in the historical test bed of the sandpile? Suppose a grain falls somewhere in the West. Where it lands the pile is already steep, and so there follows a small avalanche. The sandpile historian notes the general cause: Whenever a grain falls on a steep place it causes an avalanche. This is legitimate and perceptive. So far so good.

But suppose another grain falls elsewhere and sets off an enormous, pile-wide cataclysm. The historian can still point to the steepness of the pile where the grain fell as the cause of the avalanche. But what made this avalanche so big? To tackle this important question, the historian now has to refer to the whole complicated chain of events that carried the avalanche across the pile. As a consequence, he may find it difficult to discover a cause that can be generalized, and will only be able to talk about a string of a thousand happenings that in its details will never occur again.

Like the historians who saw the origins of the First World War in a "breakdown of the international system," the sandpile historian may attribute the events to some "peculiar and tragic configuration of the pile" that enabled a single grain to touch off disaster. And in saying this, he would be right. And yet this way of thinking is misguided if it also supposes that these "peculiar conditions" were special and unusual. In the sandpile game, as we know, such conditions are typical, and it is always possible for a grain, falling in just the right place, to trigger a terrific avalanche.

If Richardson's and Levy's power laws for the distribution of wars by size are any indication, then this may also be true of human history, as the fabric of politics both within and between nations may be in something like a critical state. On a local level, there may be easy-to-find excellent general causes, such as Read's "maladjustment" that always precedes the breakdown of a portion of the conventional and traditional fabric of any social community. Read's generalization is analogous to that of the sandpile analyst who sees slopes as the causes of sand sliding, or the forest fire analyst who understands that one burning tree can set alight another nearby. But why does one revolution or war turn out to be of no great consequence, while others spread into cataclysms? On this question, general causes may be more elusive. Even for momentous events such as world wars, historians may ultimately be able to do little more than point to origins in a nose or a monkey bite, or to the automobile driver who made a wrong turn, and then trace the chain of events that followed. Is this perhaps why they have had such difficulties agreeing on the ultimate causes for the First World War? The only general cause for such an event may be the underlying organization of the critical state, which makes such upheavals not only possible but inevitable.

If so, then the problem of Cleopatra's nose is very real, and the British philosopher Michael Oakeshott may have been right:

> . . . every historical event is necessary, and it is impossible to distinguish between the importance of necessities. No event is merely negative, none is non-contributory. To speak of a single, ill-distinguished event (for no historical event is securely distinguished from its environment) as determining, in the sense of causing and explaining, the whole subsequent course of events is . . . not bad or doubtful history, but not history at all. . . . There is no more reason to attribute a whole course of events to one antecedent event rather than another. . . . The strict conception of cause and effect appears . . . to be without relevance in historical explanation. . . .[12]

In a critical history, contingency becomes powerful beyond measure. From one perspective, wars obviously break out because men cannot seem to settle their differences without violence, and perhaps even because they have a taste for blood. But from another more abstract point of view, great wars may take place simply because the collective attitudes, ideas, and behaviors of the mass of humanity are subject to the same wild fluctuations of the magnet poised between its magnetic and nonmagnetic phases. It goes without saying that nothing I have mentioned in the past few chapters proves this. The "take-home" message is simply that this is a real possibility.

Historical Games

The typical question of history is, How have things changed? History could in principle be like the growth of a tree, and follow a simple progression toward some mature and stable end point, as both Hegel and Karl Marx thought. In this case, wars and other tumultuous social events should grow less and less frequent as humanity approaches the stable society at the End of History. Or history might be like the movement of the Moon around the Earth, and be cyclic, as the historian Arnold Toynbee once suggested. He saw the rise and fall of civilizations as a process destined to repeat itself with regularity. Some economists believe they see regular cycles in economic activity, and a few political scientists suspect that such cycles drive a correspondingly regular rhythm in the outbreak of wars. Of course, history might instead be completely random, and present no perceptible patterns whatsoever. These are the most common possibilities that historians have considered for how things can change.

But this list is incomplete. Establishing how things can change, in principle, is a matter not for history but for physics. In the 1980s, physicists discovered that even very simple things can behave in extraordinarily complicated ways. Imagine a platform that moves up and down in a perfectly regular way, like a ride at an amusement park. Suppose you drop a very bouncy rubber ball onto the platform from above, and ask the question, How high will the ball go after bouncing on the platform, say, ten times? If the platform were standing still, this would be easy: the ball would come back more or less to the height from which you dropped it. But with the platform moving, answering the question becomes practically impossible. For even the tiniest tremor in the hand that drops the ball will be amplified as the ball makes subsequent bounces, and after only a few bounces will reveal itself in a complete alteration in the ball's height. With the platform moving, the ball's trajectory becomes unstable in a profound way, so that the bounce-by-bounce record of its height looks wild and erratic. This is chaos.

It is possible, then, for something to seem almost random in its workings, and yet actually not be random at all. This is another category of change, and yet it is not quite the proper category for history. Chaos theory reveals how things that are indeed simple in their workings—such as the bouncing ball—can nevertheless appear to be very complicated. If you know the ball's height now, then it is a matter of arithmetic to work out how high it will be after one more bounce. At any moment, you need very little information to specify the state of the ball. This is a very simple game, even if it is an interesting one because it is chaotic.

History, however, is complicated. There are many people with many attitudes, many ideas, and many memories. Human history is the story of many men and women in time; it is about a collective. Hence the physics that might pertain to it is the physics of collectives. To say what is going on in an iron magnet at any moment requires not just a little information, but the information on the orientations of an astronomical number of atomic magnets. A magnet is an extraordinarily complicated thing. But if chaos teaches physicists that the truly simple can nevertheless look complicated, the critical state teaches them that the truly complicated can behave in ways that are remarkably simple. As we saw in chapter 6, the basic organization of any substance poised in the critical state between two phases depends very little on the precise nature of the elements involved. There is a profound universality at work, which makes it possible to understand literally thousands of utterly different collectives in terms of simple mathematical games that share the same skeletal logic.

Bringing the equilibrium critical state to life in a magnet requires careful tuning, and a lot of work in the laboratory. But one of the deepest discoveries in physics of the past two decades is that in nonequilibrium systems the critical state often arises on its own. If it is true that physicists are still trying to work out under just what conditions it does and does not appear, this needn't stop—and isn't stopping—scientists in other fields from taking advantage of this breakthrough. It shouldn't stop historians either.

Like chaos, the critical state bridges the conceptual gap between the regular and the random. The pattern of change to which it leads through its rise of factions and wild fluctuations is neither truly random nor easily predicted. It is a universal and understandable pattern that nonetheless slips the grasp of detailed prediction, reveals itself only in the statistics, and seems to draw the human mind into perceptual error. It does not seem normal and lawlike for long periods of calm to be suddenly and sporadically shattered by cataclysm, and yet it is. This is, it seems, the ubiquitous character of the world.

· 13 ·

Concluding
Unscientific Postscript

I will be brief. But not nearly as brief as Salvador Dali,
who gave the world's shortest speech.
He said, "I will be so brief I am already finished,"
and he sat down.

—E. O. WILSON[1]

HISTORIANS BECOME HISTORIANS BECAUSE THEY FIND HISTORY fascinating. "There is something in the nature of historical events," wrote the historian Herbert Butterfield, "which twists the course of history in a direction that no man ever intended." History *is* fascinating. But why? Why isn't history boring?

One reason, no doubt, is that the future is perpetually giving birth to true novelty. Human history is a bit like biological evolution: what there is in the present goes together in new ways to produce things in the future, the likes of which have never before existed. There are undeniable trends in history, one of the most obvious being the increase in our scientific understanding and the technological complexity of our world. Unprecedented objects, processes, and possibilities continually come into existence. But as we saw earlier, the evolutionary history of life on Earth is not interesting solely because of the novelty that has emerged in the diversity of species. The fabric of life is also fascinating because it is subject to the wild fluctuations of the critical state, revealed both in the mass extinctions and in the terrific fluctuations in the populations of species in ecosystems of all sorts.

If the same were true of human cultural evolution, and the web of connected cultures covering the planet, this might be part of the reason that history is not boring. To bring things into focus, think one more time of our magnet. Imagine that you lived in that magnetic world, and that its temperature was held below the critical point. All the magnets would be pointing in the same direction, and would rarely if ever depart from it. All your friends would be doing the same thing, and life would be one long, monotonous nothingness. The history of your world would have fixed laws and unending peace and uniformity stretching back through time: history would be truly boring. Indeed, there would be no history at all, since a record of unbroken sameness is not history, but the lack of it.

Now suppose instead that the temperature were to be raised well above the critical temperature. Now all the magnets would be flipping back and forth wildly and at random, and what any one magnet did at any moment would have no correlation whatsoever with what its neighbors were doing. In this mad world, none of your friends could be counted on for anything, there would be no shred of continuity, no semblance of order, and the past would be a wholly senseless record of absolutely random changes. Again, a boring world, since the only thing you could say of it would be, "It is random."

But how much more interesting things would be if the temperature were brought close to the critical temperature. Suddenly, any one magnet would be able to have a considerable influence on its neighbors, but not so much that they would all fall into line. Your society would now be filled with groups of all sizes, with alliances shifting constantly and in a way that would be neither quite orderly nor quite random. You would occasionally notice mass movements initiated by one or a few magnets that would unexpectedly sweep across the entire world. History would seem to be over until, after a period of calm of unpredictable length, another mass movement went the other way. No movement would be a simple copy of what had just happened, but every change would be different in its detail from what went before.

You would also find that your own activities could potentially have a great effect on what would happen in your world and that the actions of others could similarly have a profound, though not quite inescapable, influence on your own. Looking back at history, you would find a fascinating pattern that had structure and randomness all mingled in some perplexing but intoxicating way. Things would always appear as if they had settled down and were predictable, just when another great fluctuation would come along to throw the world into turmoil. History would be extremely interesting.

Does this give us a clue as to why our own human world is interesting? We have seen a few hints that our world at many levels may indeed be subject to the same kinds of fluctuations as the sandpile, or the magnet at the critical point, in which influences possess an enhanced ability "to propagate." If the social and political fabric of the world really is shaped in this way, we should learn to expect the unexpected. We live now in a time that is relatively peaceful. This relative calm may endure for another century, or we may see another world war in five years—no one can really say. The United States as a nation may survive for another five hundred years, or crumble in thirty. If the world is critical, then there are local causes that can be investigated, and we can make sense of how politics and social forces shape historical changes here and there. But if the ultimate effects of any happening depend on how the details link together to create "fingers of instability" stretching through the world, it becomes virtually impossible to see into the future. Trends cannot be trusted to continue; all that we can predict is that the future will continue to escape our grasp. History is neither static nor changing randomly, but balanced precariously in some middle way between the two, and so always poised, like the sandpile, on the very edge of dramatic upheaval.

The paleontologist and evolutionary biologist Stephen Jay Gould has rightly observed:

> We are especially moved by events that did not have to be, but that occurred for identifiable reasons subject to endless mulling and stewing. By contrast, both ends of the usual dichotomy—the inevitable and the truly random—usually make less impact on our emotions because they cannot be controlled by history's agents and objects, and are therefore either channeled or buffeted, without much hope for pushing back. But, with contingency, we are drawn in; we become involved, we share the pain of triumph or tragedy.

When we realize that the actual outcome did not have to be, that any alteration in any step along the way would have unleashed a cascade down a different channel, we grasp the causal power of individual events. We can argue, lament, or exult over each detail because each holds the power of transformation. Contingency is the affirmation of control by immediate events over destiny, the kingdom lost for want of a horseshoe nail.[2]

In our world, every tiny event is recorded, and has influence, and that influence can be world-changing. This is not necessarily the case. The world could have been organized otherwise.

Physics of History

A few historians are beginning to see that physics may indeed offer a new conceptual vocabulary well suited to their task. As the Oxford historian Niall Ferguson recently noted,

> . . . a great many of those philosophers of history who have argued in this century about whether history was a "science" seem not to have grasped that their notion of science was an out-of-date relic of the nineteenth century. What is more, if they had paid more attention to what their scientific colleagues were actually doing, they would have been surprised—perhaps even pleased—to find that they were asking the wrong question. For it is a striking feature of a great many modern developments in the natural sciences that they have been fundamentally historical in character. . . .[3]

Ferguson points to chaos theory as an important conceptual tool for the historian, one that "reconciles the notions of causality and contingency." It is true that chaos does reveal how, even for a process that is strictly deterministic, a small change at the outset can result in a very different outcome. As we have seen, however, what is missing from chaos is the essential notion of collective behavior. There are not merely a few forces at work in history, but innumerably many. To understand the typical patterns that one might expect in history, one needs the historical science of systems in which many independent things interact with one another.

This is the domain of non-equilibrium statistical physics. In such systems it is obvious that there can be no hope of making exact predictions, and yet balancing this lack of order at the level of individual events are profoundly regular and often very simple statistical laws, such as the power laws we have met over and over. These reveal the character of a deeper historical process behind specific events. It is not chaos that historians should be turning to for instruction, but universality—the notable discovery that under very broad conditions, systems made of interacting objects of all different kinds show universal features in their behavior.

In the ancient world, people typically attributed great events to the actions of the gods. In the words of one historian:

> Whenever the causes seemed incommensurate with the results or the mundane explanation seemed inadequate, whenever chance or a curious conjecture produced something that conflicted with expectations, whenever extraneous factors not normally brought into the reckoning . . . gave the narrative a surprising twist, in all these cases one would . . . believe that [God] had intervened. This recourse to divine intervention to explain the unexpected illustrates the importance of contingency in history; the inability at early stages of the development to see all the connections between the events; the cataclysmic character of the happenings; the fact that great

consequences can proceed out of little causes; the fears that men have in a world, the proceedings of which they do not understand; the feeling men have that history is a thing that happens to them rather than something that they are making; the feeling of dependence which they would doubtless have when they were unable to understand or master the operations of nature, the mystery of natural happenings . . . ; all these things would lead men to feel in life that much depended on the gods. . . .[4]

Is there a driving force?

Today, we are still baffled by great wars and revolutions, although now we confront them without the metaphysical comfort that might be afforded by the ancients' beliefs in the gods. We know that history is made by individuals acting as individuals, that the potential for both war and peace lives in every person, and that somehow, out of the mysterious ocean of individual activity, great tidal waves all too frequently rise up to sweep us away. It will make no one feel any safer or happier to realize that these waves may be inevitable. But it is at least a step toward greater understanding to recognize that the tumultuous course of humanity need not be the product of some deeply malignant human madness, but of ordinary human nature and simple mathematics.

Physics has entered a remarkable era. As the physicist James Langer of the University of California at Santa Barbara recently wrote:

> For the first time in history, we have the tools—the experimental apparatus and the computational and conceptual capabilities—to answer questions of a kind that have always caught the imagination of thoughtful human beings. . . . I cannot remember a time when I have felt more optimistic about the intellectual vitality of physics than I do now.[5]

Langer was commenting on theoretical efforts directed toward understanding a historical process: not human history, in this case, but the process of growth that leads to the infinitely intricate shapes of snowflakes. Snowflakes crystallize as ice from the thin air, and the way they grow is a perfect example of a process in which history matters. By coincidence, I am finishing off this final chapter while sitting next to a window looking out onto a mountainside in the French Alps, where a blizzard of Alpine intensity is covering the slopes with a truly fantastic number of snowflakes. The number of flakes is mind-boggling, but even more impressive is the fact that no two are identical, and that, nevertheless, they all share universal features. For several hundred years this has been a deep mystery. But it is no longer. Science has finally reached the stage at which it can make sense of the shapes of snowflakes.

Nonequilibrium physics is the new frontier for physicists, and from all the historical games in the research journals, and the remarkable insights they are now affording into hundreds of things, it is easy to understand Langer's confidence in the "intellectual vitality" of physics. This is, of course, only a beginning. Indeed, all the research I have mentioned in this book represents only the vaguest of beginnings for the science of things in which history matters, whether it is the science of earthquakes or extinctions, or the science of snowflakes, of science itself, or of human history.

In his novel *War and Peace,* Tolstoy asked: "Why do wars and revolutions happen?" Perhaps it is not too much to suppose that what physicists are learning in their own special way, with their penchant for oversimplification and abstraction, may one day permit historians to improve on his answer:

> We do not know. We only know that to produce the one
> or the other men form themselves into a certain combi-
> nation in which all take part; and we say that this is the
> nature of men, that this is a law.

Notes and References

✦

1. Causa Prima

1. John Kenneth Galbraith, letter to John F. Kennedy, March 2, 1962, *Ambassador's Journal* (Boston: Houghton Mifflin, 1969), p. 312.

2. Paul Valéry, *Variété IV* (Paris: Gallimard, 1938).

3. A.J.P. Taylor, *The First World War* (London: Penguin, 1963).

4. See, for example, Niall Ferguson, *The Pity of War* (London: Penguin, 1998).

5. Clarence Alvord, quoted in Peter Novick, *That Noble Dream* (Cambridge: Cambridge University Press, 1988), pp. 131–132.

6. Ferguson, *Pity of War*, p. 146.

7. Francis Fukayama, *The End of History and the Last Man* (New York: Free Press, 1992).

8. H.A.L. Fisher, quoted in Richard Evans, *In Defence of History* (London: Granta Books, 1997), pp. 29–30.

9. The City of Kōbe website, http://www.city.kobe.jp/.

10. Paul Somerville, "The Kōbe Earthquake: An Urban Disaster," *Eos* 76 (1995): 49–51.

11. Quoted in Rocky Barker, *Yellowstone Fires and Their Legacy*, available on-line at http://www.idahonews.com/yellowst/yelofire.htm.

12. *Wall Street Journal*, September 23, 1987.

13. *Wall Street Journal*, August 26, 1987.

14. *Wall Street Journal*, October 7, 1987.

15. Robert Prechter Jr., *The Wave Principle of Human Social Behaviour* (Gainesville, GA: New Classics Library, 1999), p. 378.

16. William James, *Principles of Psychology* (1890; Dover, 1950), vol. 2, chap. 22.

17. Albert Camus, *The Myth of Sisyphus*, trans. Justin O'Brien (London: Penguin, 1975).

18. Per Bak, Chao Tang, and Kurt Weisenfeld, "Self-Organised Criticality: An Explanation of 1/f Noise, *Physical Review Letters* 59 (1987): pp. 381–384.

19. Per Bak, *How Nature Works* (Oxford: Oxford University Press, 1996).

20. As it turns out, one thing Bak, Tang, and Weisenfeld's computer sandpile game *doesn't* mimic accurately is avalanching in a real sandpile. But that is of little consequence; ironically, their computer game now seems immeasurably more important than any real pile as an exemplar of a particularly fascinating way that our world works at many levels. More on this in chapter 7.

21. Some readers may wonder what the critical state has to do with "catastrophe theory," another idea of recent years that, based on its name, may seem quite similar. In truth, there is very little connection. If you push very gently on both ends of a drinking straw, as if to try to compress it and make it shorter, it will indeed become ever so slightly shorter. Push harder, however, and there will come a point when the straw will abruptly and suddenly bend. In the 1970s, a mathematician named René Thom worked out a theory to make sense of sudden changes in simple cases of this sort, changes that he referred to as "catastrophes." But Thom's Catastrophe Theory, despite its provocative name, has very little to say about the workings of anything like the Earth's crust or an economy or an ecosystem. In these things, where thousands or millions of elements interact, what is important is the overall *collective* organization. To understand things of this sort, one needs a theory that applies generally to networks of interacting things, something for which Catastrophe Theory is ill prepared.

22. Paul Kennedy, *The Rise and Fall of the Great Powers* (New York: Random House, 1987).

2. A Shaky Game

1. Paul Valéry, *Collected Works*, vol. 14, *Analects*, ed J. Matthew (London: Routledge, 1970).

2. Charles Richter, acceptance of the Medal of the Seismological Society of America, *Bulletin of the Seismological Society of America* 67 (1977): 1244–1247.

3. See the comments of Ian Main, opening a *Nature* on-line debate on the topic "Is Earthquake Prediction Possible?" Organized in the spring of 1999, this debate involved many of the world's leading earthquake experts. Contributions can be found at http://www.nature.com/.

4. W. Spence, R. B. Hermann, A. C. Johnston, and G. Reagor, "Responses to Iben Browning's Prediction of a 1990 New Madrid, Missouri, Earthquake," *U.S. Geological Survey Circular* 1083 (U.S. Government Printing Office, 1993).

5. R. Rikitake, "The Large-Scale Earthquake Countermeasures Act and the Earthquake Prediction Council in Japan," *Eos, Transactions, American Geophysical Union* 60 (1979): 553–555.

6. B. T. Brady, "Theory of Earthquakes," IV: "General Implications for Earthquake Prediction," *Pure and Applied Geophysics* 114 (1976): 1031–1082.

7. The earthquake magnitude, which reflects this total energy, works logarithmically. So a magnitude 7 quake is ten times more powerful than one with magnitude 6.

8. W. H. Bakun and A. G. Lindh, "The Parkfield, California, Earthquake Prediction Experiment," *Science* 229 (1985): 619–624.

9. C. F. Shearer, "Minutes of the National Earthquake Prediction Evaluation Council, March 29–30, 1985," *USGS Open File Report* 85–507.

10. "A Proposed Initiative for Capitalizing on the Parkfield, California Earthquake Prediction," *Commission on Physical Sciences, Mathematics, and Resources*, National Research Council (Washington D.C.: National Academy Press, 1986).

11. "Small Earthquake Somewhere, Next Year Perhaps," *Economist*, August 1, 1987.

12. Y. Y. Kagan, "Statistical Aspects of Parkfield Earthquake Sequence and Parkfield Prediction Experiment," *Tectonophysics* 270 (1997): 207–219.

13. Mark Twain, *Life on the Mississippi*.

14. R. J. Geller, "Predictable Publicity," *Seismological Research Letters* 68 (1997): 477–480.

15. A much less detailed kind of "prediction" is of course used all the time. We know that many earthquakes have struck in the past in places such as California and Japan, while not many have struck in New York State or in Britain. Consequently, we recognize that the risk of earthquakes is higher in some places than in others. This knowledge quite usefully informs building codes in high-risk areas, but still doesn't give any knowledge about specific earthquakes.

16. R. J. Geller, "Earthquake Prediction: A Critical Review," *Geophysical Journal International* 131 (1997): 425–450.

17. Evans, *In Defence of History*, p. 59.

18. To be precise, the stuff of the mantle isn't actually liquid, but solid. It is a collection of minerals. And yet it is so hot and under such tremendous pressure that it still manages to flow about as if it were liquid. The movement is just very slow.

19. Eliza Bryan, in *Lorenzo Dow's Journal* (Joshua Martin, 1849), pp. 344–346.

20. The website is http://www-socal.wr.usgs.gov/index.html.

21. There is one exception. So-called deep earthquakes don't appear to occur in the Earth's brittle crust, but far below, when a region of rock under tremendous pressure suddenly undergoes a "change of phase"—that is, the way its molecules are packed together suddenly changes. This results in an abrupt change in the volume of the rock, triggering an earthquake.

3. An Absurd Reasoning

1. Friedrich Nietzsche, *Twilight of the Idols; and, the Anti-Christ*, tr. R. J. Hollingdale (New York: Penguin Books, 1990), p. 62.

Notes and References

2. Eric Temple Bell, *Mathematics: Queen and Servant of Science* (New York: McGraw-Hill, 1951).

3. Camus, *Myth of Sisyphus.*

4. If you are wondering where the term "power law" comes from, the answer is as follows. In algebra, a power law is any curve for which the height changes in proportion to the horizontal distance raised to some power—that is, multiplied by itself a certain number of times. For example, the equation height = (distance)^2 represents a curve that bends upward ever more steeply. This is a power law with power equal to 2. In the case of earthquakes, if we think in terms of energy rather than magnitude, the Gutenberg-Richter curve says that the number of earthquakes having energy equal to some value E is inversely proportional to E raised to the power two, or E^2. This power law captures the simple pattern we've been talking about: each time you double the energy, the earthquakes being considered become four (that is, two squared) times as rare.

5. R. Burridge and L. Knopoff, *Bulletin of the Seismological Society of America* 57 (1967): 341.

6. P. Bak and C. Tang, "Earthquakes as a Self-Organized Critical Phenomena," *Journal of Geophysical Research B* 94 (1989): 15635.

7. I should mention that Bak and Tang were not alone in making the link between sandpiles and earthquakes. Around the same time, several other researchers simultaneously and independently came to similar conclusions. See, for instance, A. Sornette and D. Sornette, "Self-Organized Criticality and Earthquakes," *Europhysics Letters* 9 (1989): 197; and K. Ito and M. Matsuzaki, "Earthquakes as Self-Organized Critical Phenomena, *Journal of Geophysical Research B* 95 (1990): 6853.

8. Z. Olami, H. J. Feder, and K. Christensen, "Self-Organized Criticality in a Continuous, Non-Conservative Cellular Automaton Modeling Earthquakes," *Physical Review Letters* 68 (1992): 1244–1247.

9. K. Ito, "Punctuated Equilibrium Model of Evolution Is Also an SOC Model of Earthquakes," *Physical Review E* 52, (1995): 3232–3233.

10. The mathematical signature of clustering is the distribution of "waiting times" for the next large earthquake, after one large quake has taken place, a distribution that also follows a power law. For a huge number of earthquakes in the game, Ito recorded the time until the next quake struck, finding shorter times more frequently than longer, and in the usual power law style: waiting times of, say, two weeks turn up about 2.8 times less frequently than times of just one week, and the same goes for times of two months relative to one month, two years relative to one year, and so on. Compare this to the same distribution for real quakes, which is known as the Omori law: in the real world, the number in the power law for waiting times isn't 2.8, but is awfully close, being about 2.6.

4. Critical Thinking

1. Quoted in Alan L. Mackay, comp., *A Dictionary of Scientific Quotations* (Philadelphia: A. Hilger, 1991).

2. Samuel Karlin, Eleventh R. A. Fischer Memorial Lecture, Royal Society, April 20, 1983.

3. Hendrik Jensen, *Self-Organized Criticality*, Cambridge Lecture Series in Physics, 10 (Cambridge: Cambridge University Press, 1998), p. 148.

4. B. Malamud, G. Morein, and D. Turcotte, "Forest Fires: An Example of Self-Organized Critical Behaviour," *Science* 281 (1998): 1840–1842.

5. Stephen Pyne, *America's Fires* (Durham, N.C.: Forest History Society, 1997).

6. Steve Allison-Bunnell, "The Dance of Life and Death," a series of articles about forest fires available on the Web at http://www.discovery.com/.

7. Federal Wildland Fire Policy, available on the Web at http://www.fs.fed.us/.

8. D. Lockwood and J. Lockwood, "Evidence of Self-Organized Criticality in Insect Populations," *Complexity* 2 (1999): 49–58.

9. C. J. Rhodes and R. M. Anderson, "Power Laws Governing Epidemics in Isolated Populations," *Nature* 381 (1996): 600–602.

10. R. Garcia-Pelayo and P. D. Morley, "Scaling Law for Pulsar Glitches," *Europhysics Letters* 23 (1993): 185.

11. V. Frette et al., "Avalanche Dynamics in a Pile of Rice," *Nature* 379 (1996): 49–52.

12. Bak, *How Nature Works*, p. 51.

13. See, for example, Alessandro Vespignani and Stefano Zapperi, "How Self-Organized Criticality Works: A Unified Mean-Field Picture," *Physical Review E* 57 (1998): 6345–6362. See also Ronald Dickman, Miguel Muñoz, Alessandro Vespignani, and Stefano Zapperi, "Paths to Self-Organized Criticality," Los Alamos e-print (cond-mat/9910454).

5. Killing Time

1. Laurence Sterne, *Tristram Shandy* (London: Wordsworth Editions, 1996).

2. Daniel Dennett, *Darwin's Dangerous Idea* (London: Allen Lane, 1995), p. 21.

3. A nice account of what it is like to look for fossils in eastern Montana is Peter Ward's *The End of Evolution* (London: Weidenfeld & Nicholson, 1995).

4. Perhaps not quite all. Palaeontologists now believe that a few dinosaurs made it through, and subsequently evolved into the modern birds. So there are still dinosaurs on Earth after all!

5. F. B. Loomis, "Momentum in Variation," *American Naturalist* 39 (1905): 839–843.

6. F. Nopsca, "Notes on British Dinosaurs," IV: *"Stegosaurus priscus,"* *Geological Magazine* 8, (1911): 143–153.

7. Michael Benton, "Scientific Methodologies in Collision: A History of the Study of the Extinction of the Dinosaurs," *Evolutionary Biology* 24 (1990): 371–400.

8. This reminds me of a clever bit of psychology (and clear thinking) employed by John Maddox, the former editor of *Nature.* Maddox had little sympathy for authors who wanted to title their paper "Evidence for . . . ," and always insisted that a paper's title should describe the facts that the work really established, rather than what those facts might possibly be taken to imply. If authors objected, as they often did, Maddox offered to leave "Evidence for" in the title, so long as it was, for clarity, modified to "Inconclusive Evidence for." I don't believe there were any takers.

9. S. A. Bowring et al., "U/Pb Zircon Geochronology and Tempo of the End-Permian Mass Extinction," *Science* 280 (1998): 1039–1045.

10. Charles Darwin, *Origin of Species by Means of Natural Selection* (1859; New York: Penguin, 1985), p. 321. This isn't to disparage Darwin. If today we have the luxury of arguing about the subtler rhythms of evolution, Darwin was waging a battle to convince his contemporaries that evolution was a reality at all. As Steven Stanley points out, Darwin had to assure his readers that evolution operates with extreme slowness if only to disarm those who would "argue against his theory simply because, while they could witness artificial selection in the barnyard, they could not observe selection in nature." See Steven Stanley, *Macroevolution* (San Francisco: W. H. Freeman, 1979; Baltimore: Johns Hopkins University Press, 1998).

11. L. W. Alvarez, W. Alvarez, F. Asaro, and H. V. Michel, "Extraterrestrial Cause for the Cretaceous-Tertiary Extinction," *Science* 208 (1980): 1095–1108.

12. Walter Alvarez, *T. rex and the Crater of Doom* (Princeton, NJ: Princeton University Press, 1997), p. 15.

13. Ibid.

14. Ironically, the crater had actually been discovered only one year after Alvarez and his colleagues published their first paper on the catastrophic extinction by impact. Researchers engaged in oil exploration had in 1981 mapped out the site and identified it as the world's largest impact crater. But they didn't know about the Alvarez idea. It took another ten years for scientists to make the connection.

15. Benton, "Scientific Methodologies in Collision."

16. Alvarez, *T. Rex and the Crater of Doom,* p. 15.

17. K. A. Farley, A. Montanari, E. M. Shoemaker, and C. S. Shoemaker, "Geochemical Evidence for a Comet Shower in the Late Eocene," *Science* 280 (1999): 1250–1253.

18. Steven M. Stanley, *Extinction* (New York: Scientific American Library, dist. by W. H. Freeman, 1987), p. 40.

19. Paul Wignall, *New Scientist,* January 25, 1992, p. 55.

20. David M. Raup, *Extinction: Bad Genes or Bad Luck* (New York: W. W. Norton, 1991), pp. 112–113.

21. David Jablonski, "Background and Mass Extinctions: The Alternation of Macro-Evolutionary Regimes," *Science* 231 (1986): 131.

22. Richard Leakey and Roger Lewin, *The Sixth Extinction* (London: Weidenfeld & Nicholson, 1996), p. 62.

23. J. J. Sepkoski, "Ten Years in the Library: New Data Confirm Palaeontological Patterns," *Palaeobiology* 19 (1993): 43.

24. Michael J. Benton, "Diversification and Extinction in the History of Life," *Science* 268 (1995): 52–58.

25. I am indebted to a review article on the topic, M.E.J. Newman and R. G. Palmer, "Theoretical Models of Extinction: A Review," Santa Fe Institute Working Paper 99-08-061 (1999).

6. The Web of Life

1. Umberto Eco, *Serendipity* (London: Weidenfeld & Nicholson, 1999), p. 21.

2. P. Yodzis, "The Indeterminacy of Ecological Interactions, as Perceived through Perturbation Experiments," *Ecology* 69 (1988): 508–515.

3. In this regard, Charles Darwin once speculated that "the presence of a feline animal in large numbers in a district might determine . . . the frequency of certain flowers in that district." After all, mice like to raid bumblebee nests for food, so more mice means fewer bumblebees. Cats like to eat mice, so more cats means less mice and more bumblebees. And since bees pollinate red clover and purple and gold pansies, more bumblebees means more flowers. The English love of house cats should, by this unexpected but fairly direct path, lead to prettier gardens. See Jocelyn Kaiser, "Of Mice and Moths and Lyme Disease?," *Science* 279 (1998): 984.

4. S. Kauffman and S. Johnsen, "Coevolution to the Edge of Chaos— Coupled Fitness Landscapes, Poised Sets, and Coevolutionary Avalanches," *Journal of Theoretical Biology* 149 (1991): 467. Also: Stuart A. Kauffman, *Origins of Order* (New York: Oxford University Press, 1993).

5. For example, suppose that each rare genetic event has a 1 in 100 chance of happening over a year's time. Then the chance of getting three in rapid succession is something like 1 in 1,000,000, and the chance of hitting on ten in a row is $1/10^{20}$. The likelihood of making a long jump diminishes very rapidly as the jumping distance grows.

6. In writing about the model, Bak has often slipped into speaking about the "fitness of a species." To simplify the discussion, he has spoken of the lengths of the sticks as representing a "species fitness," rather than the length of a leap a species has to make if it is to evolve further. This is a red rag to the biological bull, as

fitness—at least for most biologists—isn't a property that can properly be assigned to species. The orthodox (though controversial) view is that fitness is properly attributed only to individuals, since evolution works at the level of the individual. However, it makes no sense to criticize the Bak-Sneppen game on this point, as in its most legitimate form it makes no reference whatsoever to the fitness of species.

7. It is ironic that biologists have reacted so vociferously to the Bak-Sneppen game, since it appears to be a mathematically plausible rendering of a game once dreamed up originally by Charles Darwin himself. "The Face of Nature," Darwin wrote in *The Origin of Species*, "may be compared to a yielding surface, with ten thousand sharp wedges packed close together and driven inwards by incessant blows, sometimes one wedge being struck, and then another. . . ." Imagine, with Darwin, a number of wedges driven upward into a wooden ceiling. Each wedge represents a species, and the depth of that wedge its degree of adaptation. Wood isn't a perfectly orderly material, so each wedge will be stuck in place by a different "pinning force." Some wedges will be more firmly fixed and harder to hammer in farther than others. Now suppose that whoever does the hammering moves about at random, tapping the wedges first weakly and then more strongly until finally he or she hits a wedge and it moves. Most likely, this will be the wedge held in place by the weakest pinning force. After the blow, it will be buried more deeply, and pinned by a new force, which might be greater or lesser than before. The wielder of the hammer starts again. This reproduces the Bak-Sneppen game exactly, if we make the one additional assumption that when a wedge is hammered in, its movement also alters the pinning force on its neighboring wedges—certainly what Darwin had in mind, and quite a realistic assumption since the presence of a wedge alters the stress in the wood in its vicinity.

8. Paol Anderson, *New Scientist*, September 25, 1969, p. 638.

9. M.E.J. Newman, "Self-Organized Criticality, Evolution, and the Fossil Extinction Record," *Proceedings of the Royal Society B* 263 (1996): 1605–1610.

10. Francis Crick, *What Mad Pursuit* (London: Weidenfeld & Nicholson, 1988), p. 136.

11. Newton was, of course, simply trying to work out the basic laws for motion. Had he been managing a project to send a rocket to the moon, he would have needed to take more details into account.

7. *De Magnete*

1. J. Robert Oppenheimer, *The Open Mind* (New York: Simon & Schuster, 1955).

2. Homer Adkins, *Nature* 312 (1984), p. 212.

3. In Mackay, *Dictionary of Scientific Quotations.*

4. A superfluid lacks any trace of what every ordinary liquid has: viscosity, a kind of internal friction that ultimately puts the brakes on any motion. Honey has

a great deal of viscosity; water not so much; and a superfluid none at all. This also implies that if you begin spinning the bowl in which a superfluid resides, the superfluid will stay put. A few years ago physicist Richard Packard and his colleagues at the University of California at Berkeley exploited this property in a beautiful experiment. They put superfluid into a tiny circular channel, something like a miniature pig's trough bent round and made to connect with itself. Then they set this ring on a bench in the lab, so that, being connected to the earth, it rotated once a day. To stay put, the superfluid had to flow through the ring. By measuring the flow, Packard's team could measure the rate of the earth's rotation to within one part in a thousand.

5. Dennett, *Darwin's Dangerous Idea*, p. 174.

6. Images adapted from J.J. Binney et al., *An Introduction to the Theory of Critical Phenomena* (Oxford: Oxford University Press, 1992).

7. It is purely a matter of chance that the magnets are down rather than up. In a thousand simulations, the magnets would fall approximately five hundred times into the up state, and approximately five hundred into the down. The magnets don't have any preference for one over the other. Which one they end up in depends on accidental details of the random state from which the magnets begin. This is a case of what physicists refer to as spontaneous symmetry breaking. Even though the original problem is symmetric in that up and down are physically equivalent, what happens in any specific instance can break this symmetry.

8. To make things as simple as possible, I am departing slightly from the usual language of physics. Physicists don't talk about "critical numbers" but about "critical exponents." The relationship between the two is very simple. The Gutenberg-Richter power law says that the number of earthquakes releasing energy E is inversely proportional to E^2. So, double the energy of the earthquake, and it becomes four times as rare. Throughout this book, I am specifying the precise nature of this or that power law by using numbers such as this "four." Double the size of the thing in question, and how much less frequent does it become? This is my critical number. The critical exponent, on the other hand, is the "two" that appears as the power (exponent) to which E is raised in the Gutenberg-Richter law. So, the relationship between the two numbers is this: My critical number is equal to two raised to the power of the critical exponent, or $2^{(\text{critical exponent})}$. This may seem like a strange contortion; I have done it in order to avoid introducing the idea of a number being raised to a power that is not a whole number, such as 1.5 or 1.31 or even minus 1.6, which seems to confuse some people. In any event, there is absolutely nothing sacred about which number one uses to specify a power law. The important point is that there are different power laws, and yet all of them share the same special, self-similar character.

9. This statement applies to what physicists refer to as continuous phase transitions, excellent examples of which are the magnetic phase transition in iron or the superfluid transition in liquid helium. There are thousands of other examples.

But when ordinary water boils and turns to vapor, this transition falls into a somewhat different class, as it does not take place by way of a critical state. However, when water is heated and turned to vapor at a particular pressure (a quite high pressure, as it turns out), then the liquid–vapor transition does indeed take place via a critical state. This is true for all the liquid–vapor transitions I have mentioned. This point is of absolutely no importance for conclusions of this chapter. I include it only to point out that the field of phase transitions is rather deeper than my words seem to suggest.

10. Almost nothing else. In physics, there is always an exception. One other thing that matters is the range of the interaction between particles. If particles are able to interact over long distances, this can push the system from one universality class to another.

11. Mathematical proof arrived in the 1970s in the form of something called the theory of the Renormalization Group, initiated by Kenneth Wilson of Cornell University, and for which he eventually won the Nobel Prize. One might say that the Renormalization Group *proved* the principle of universality.

12. In 1995, for example, physicists from the ETH in Zurich, Switzerland, made a very thin magnet. They laid down a film of iron one atom thick, and then deposited some more iron atoms in an irregular array of patches. In a few places, they put down a third layer by accident. The result was a hideously poor realization of Onsager's "flatland" magnet. The iron layer was not perfectly two-dimensional, nor were the magnets arranged into a perfectly regular lattice. There were also other differences as well. Because iron is a metal, the laws of quantum theory imply that its atomic magnets aren't in fact even localized; instead, each is in some completely strange way "spread out" over a large region. What's more, whereas the interactions between neighboring atoms in the toy are all exactly the same, in the irregular iron film they are all slightly different. Even so, the critical exponents for this crude thin slab of iron agreed *perfectly* with those of Onsager's model. The dimension and the dimension of the order parameter were the same, and that was enough. C. H. Back et al., "Experimental Confirmation of Universality for a Phase Transition in Two Dimensions," *Nature* 378 (1995): 597–600.

13. The critical state that we have met in this chapter arises in equilibrium systems when they are poised between two different phases. For equilibrium systems, physicists have a general recipe for working out how the system should behave at any temperature, and critical-state universality follows from it by way of Kenneth Wilson's Renormalization Group ideas. For things that are out of equilibrium, no one has yet discovered any general recipes. As a result, there is no strict theorem of universality yet for nonequilibrium systems. Nevertheless, physicists have found that many of the simple nonequilibrium games they have studied do indeed fall into universality classes. So it appears that some kind of universality will hold for nonequilibrium systems as well.

One other point: Mark Newman's game for extinctions organizes itself into a state that, technically speaking, isn't a critical state. The species in the game do not interact with one another in any way, so there can be no "rise of factions"—that is, no correlations in the viabilities of different species, and no chains of instability running through the system. Nevertheless, the game does show power laws and the same wild sensitivity in which a quite ordinary, run-of-the-mill shock can trigger an upheaval out of all proportion to itself. What's more, as with the critical state, this sensitivity comes about naturally under an extremely broad set of conditions. These are the ubiquitous properties that arise again and again in things driven away from equilibrium, and in things in which history matters. Since Newman's game shares the spirit of the critical state, I have spoken of it as if it were indeed in such a state.

8. Wild at Heart

1. Wassily Leontief, letter to the editor, *Science*, July 9, 1982.
2. Alfred Zauberman, *Guardian*, October 5, 1983.
3. John Kay, "Cracks in the Crystal Ball," *Financial Times*, September 29, 1995.
4. OECD Economic Outlook, June 1993.
5. John Rothchild, *The Bear Book* (New York: John Wiley, 1998).
6. Then again, even making predictions of this simple and obvious sort may not be so easy. A 1997 study asked five of the U.K.'s top orthodox economic modeling groups to say how the economy would respond to an incremental increase in public spending. The results point up the lack of theoretical consensus even on seemingly simple issues such as this. Not only did the different groups find different numbers, they couldn't even agree on whether there would be an overall increase or decrease in economic output. See Paul Ormerod, *Butterfly Economics* (London: Faber & Faber, 1998).
7. Rudiger Dornbush, "Growth Forever," *Wall Street Journal*, July 30, 1998.
8. Robert Schiller, quoted in Robert Prechter Jr., *The Wave Principle of Human Social Behaviour* (Gainesville, GA: New Classics Library, 1999).
9. If orthodox economic theory claims that prediction is possible, and yet fails so miserably, why isn't it dismissed as being useless? One analyst suggests, not implausibly, that economists don't reject their favorite theoretical relationships when they fail because that would "destroy the utility of the tool. It is easier for the neocortex to rationalise when it does not know or think too much. That is why one day in a bull market it can be found engaged in explaining why a rise in the Nikkei is bullish for US stocks ('It is a vote of confidence that the Japanese recession will not deepen and so will not spread to the US'), and the next day it can be found engaged in explaining why a drop in the Nikkei is bullish for US stocks ('That money will have to flee to a stronger market')." Prechter, *Wave Principle of Human Social Behaviour.*

Notes and References

This chapter draws quite heavily on Prechter's devastating critique of contemporary economic thought.

10. J. D. Farmer and A. Lo, "Frontiers of Finance: Evolution and Efficient Markets," Santa Fe Institute Working Paper 99-06-039 (1999).

11. Alan Kirman, quoted in Ormerod, *Butterfly Economics*, p. 16.

12. R. E. Litton and A. M. Santomero, *Wall Street Journal*, July 28, 1998.

13. John Casti, "Flight over Wall Street," *New Scientist*, April 19, 1997.

14. Benoit Mandelbrot, *Journal of Business* 36 (1963): 294.

15. P. Gopikrishnan, M. Meyer, L.A.N. Amaral, and H. E. Stanley, "Inverse Cubic Law for the Distribution of Stock Price Variations," *European Physical Journal B* 3, 139 (1998).

16. Vasiliki Plerou, et al., "Scaling of the Distribution of Price Fluctuations of Individual Companies," Los Alamos e-print (cond-mat/9907161), July 11, 1999. To be technically accurate, I shouldn't speak of price fluctuations but of what is known as market capitalization. The S&P 500 index isn't just an average of the share prices of the stocks of five hundred companies, but a weighted average, in which the companies with more outstanding shares count more heavily. Similarly, research on single companies looks at the fluctuations in a quantity that is the share price multiplied by the number of outstanding shares. None of this alters the conclusion that the market has large fluctuations.

17. R. N. Mantegna, "Levy Walks and Enhanced Diffusion in the Milan Stock Exchange," *Physica A* 179 (1991): 232.

18. Pictet, O.V. et al., "Statistical Study of Foreign Exchange Rates, Empirical Evidence of a Price Change Scaling Law and Intraday Analysis," *Journal of Banking and Finance* 14 (1995): 1189–1208.

19. Liu, et al. "Statistical Properties of the Volatility of Price Fluctuations," *Physical Review E* 60 (1999): 1–11.

20. Ormerod, *Butterfly Economics*, p. 36.

21. D. Sornette and D. Zajdenweber, "Economic Returns of Research: the Pareto Law and Its Implications," Los Alamos e-print (cond-mat/9809366), September 27, 1998.

22. Thomas Lux and Michele Marchesi, "Scaling and Criticality in a Stochastic Multi-agent Model of a Financial Market," *Nature* 397 (1999): 498–500.

23. Bernard Baruch, quoted in Prechter, *Wave Principle of Human Social Behaviour.*

24. Duncan Watts and Steve Strogatz, "Collective Dynamics of 'Small-World' Networks," *Nature* 393 (1998): 440–442.

9. Against All Will

1. André Gide, *The Immoralist*, trans. Richrad Howard (New York: Alfred A. Knopf: 1970), p. 7.

2. Fyodor Dostoyevsky, *Notes from Underground*, trans. Jessie Coulson (Baltimore: Penguin, 1972), p. 41.

3. D. Helbing, J. Keltsch, and P. Molnar, "Modelling the Evolution of Human Trail Systems," *Nature* 388 (1997): 47–50.

4. D. Zanette and S. Manrubia, "Role of Intermittency in Urban Development: A Model of Large-Scale City Formation," *Physical Review Letters* 79 (1997): 523–526.

5. See H. A. Makse, S. Havlin, and H. E. Stanley, "Modelling Urban Growth Patterns, *Nature* 377 (1995): 608–612; also Michael Batty and Paul Longley, *Fractal Cities* (San Diego: Academic Press, 1994).

6. J.-P. Bouchaud and M. Mézard, "Wealth Condensation in a Simple Model of the Economy," Los Alamos e-print (cond-mat/0002374), February 24, 2000.

10. Intellectual Earthquakes

1. Evans, *In Defence of History*, p. 61.

2. J. F. Jameson, quoted in Peter Novick, *That Noble Dream.*

3. Sidney Bradshaw Fay, *The Origins of the World War*, 2d rev. ed. (New York: Macmillan, 1930).

4. Charles Beard, "Heroes and Villains of the World War," *Current History* 24 (1926).

5. Harry Elmer Barnes, *The Genesis of the World War: An Introduction to the Problem of War Guilt* (New York: Alfred A. Knopf, 1926), pp. 658–59.

6. In Evans, *In Defence of History.*

7. Edward Hallett Carr, *What Is History?* (New York: Penguin, 1987), p. 9.

8. Carr, *What Is History?*

9. Conyers Read, quoted in Novick, *That Noble Dream*, p. 192.

10. Thomas Carlyle, quoted in Carr, *What Is History?*

11. Indeed, when Conyers Read made the comments quoted above, he was referring only obliquely to crises of national or international politics. He wrote those words in 1937, when concerned with a crisis in the American historical profession, which was then under considerable pressure to adapt to educationists' demands that social studies be included in school curricula. While the most conservative of historians were incensed by the idea, Read counseled that digging in their heels would only create the conditions for a revolutionary change of even greater scope.

12. Novick, *That Noble Dream*, p. 192.

13. Michael Polanyi, "The Potential Theory of Adsorption: Authority in

Science Has its Uses and Its Dangers," *Science* 141 (1963): 1012. I've slightly altered the order of words in the middle sentence for clarity.

14. Thomas Kuhn, *The Structure of Scientific Revolutions*, 3d ed. (Chicago: University of Chicago Press, 1996), p. 10.

15. Kuhn was often careful to say that a paradigm involves not only theoretical ideas but a swarm of other implicit and explicit notions and practices, all of which contribute to the application of these ideas to nature. So it is not strictly correct to say that a paradigm is a "bundle of Good Ideas." It makes the discussion a lot simpler, however, and none of these qualifications will be important in what follows.

16. Kuhn, *Structure of Scientific Revolutions*, p. 24.

17. In Ralph Kronig, "The Turning Point," in *Theoretical Physics in the Twentieth Century: A Memorial Volume to Wolfgang Pauli*, ed. M. Fierz and V. F. Weisskopf (New York: Interscience, 1960), p. 22.

18. Kuhn, *Structure of Scientific Revolutions*, p. 6.

19. In Kronig, "Turning Point," pp. 25–26.

20. Novick, *That Noble Dream*, p. 526.

11. A Question of Numbers

1. John Krasher Price, *Government and Society* (New York: New York University Press, 1954).

2. Carr, "Notes for a Second Edition," *What Is History?*

3. Kuhn, *Structure of Scientific Revolutions*, p. xi.

4. Sidney Redner, "How Popular Is Your Paper?" *European Physical Journal B* (1998): 131–134.

5. Quoted in Cyril Domb, *The Critical Point* (Bristol, Pa.: Taylor & Francis, 1996), p. 130.

6. In Mackay, *Dictionary of Scientific Quotations*.

7. Kennedy, *Rise and Fall of the Great Powers*, p. xvi.

8. Kuhn, *Structure of Scientific Revolutions*, p. 92.

9. Dostoyevsky, *Notes from Underground*.

10. Jack S. Levy, *War in the Modern Great Power System, 1495–1975* (Lexington: University of Kentucky Press, 1983), p. 215. I thank Bruce Malamud for pointing out Levy's work to me.

11. D. L. Turcotte, "Self-Organized Criticality," *Reports on Progress Physics* 62 (1999): 1377–1429.

12. Norman Davies, *Europe* (Pimlico, 1997), p. 900.

13. Carr, *What Is History?* p. 52.

14. Quoted in Novick, *That Noble Dream*, p. 139.

12. History Matters

1. Evans, *In Defence of History*, p. 62.
2. In Mackay, *Dictionary of Scientific Quotations*.
3. Carr, *What Is History?* p. 54.
4. Quoted in Carr, *What Is History?* p. 47.
5. Carr, *What Is History?* p. 46.
6. Carr, *What Is History?* p. 49.
7. A. de Tocqueville, *Democracy in America* (New York: Vintage Books, 1945).
8. Evans, *In Defence of History*, p. 133.
9. Georg Wilhelm Friedrich Hegel, *Philosophy of Right*, trans. T. M. Knox (London: Oxford University Press, 1987) p. 295.
10. Quoted in Evans, *In Defence of History*, p. 62.
11. Carr, *What Is History?* p. 23.
12. Quoted in Niall Ferguson, "Virtual History: Towards a 'Chaotic' Theory of the Past," in *Virtual History*, ed. Niall Ferguson (London: Picador, 1997), p. 50.

13. Concluding Unscientific Postscript

1. Commencement address at Pennsylvania State University, quoted in Duncan Watts, *Small Worlds* (Princeton: Princeton University Press, 1999).
2. Stephen Jay Gould, *Wonderful Life* (New York: W. W. Norton, 1989; Penguin, 1991), p. 284.
3. Ferguson, "Virtual History," p. 72.
4. Herbert Butterfield, *The Origins of History* (New York: Basic Books, 1981), pp. 200–201. See Ferguson, "Virtual History," p. 20.
5. J. S. Langer, "Nonequilibrium Physics," in *Critical Problems in Physics* (Princeton: Princeton University Press, 1997).

Index

✦

530.01   Buchanan, Mark.
BUC
           Ubiquity.

$24.00

DATE			